A HISTORY OF
ARTILLERY

A HISTORY OF
ARTILLERY

Colonel H. C. B. Rogers, O. B. E.

THE CITADEL PRESS
SECAUCUS, NEW JERSEY

First American edition, 1975
Copyright © by H. C. B. Rogers
All rights reserved
Published by Citadel Press
A division of Lyle Stuart, Inc.
120 Enterprise Ave., Secaucus, N. J. 07094

Originally published in the United Kingdom
as *Artillery Through the Ages*

Manufactured in the United States of America
Library of Congress catalog card number: 74-80824
ISBN 0-8065-0441-2

To

My Wife
who in 1940
prayed to St Barbara
to direct the thunder of the guns

CONTENTS

THE PLATES

Illustrations 21–28 inclusive are from the *Treatise on Military Carriages*, 1911.

LINE ILLUSTRATIONS

ACKNOWLEDGEMENTS

I am indebted to my friend Lieutenant-General Sir George Cole K.C.B., C.B.E., Colonel Commandant of the Royal Artillery, for his help and encouragement in writing this book. I owe much to the kindness of the late General Sir Robert Mansergh G.C.B., K.B.E., M.C., Master Gunner, St. James's Park in allowing me access to the library and records of the Royal Artillery Institution. Major-General B. P. Hughes, C.B., C.B.E., Chairman of the Royal Artillery Institution, gave me invaluable advice and took great pains to ensure that everything I required should be placed at my disposal; and the Assistant Secretary of the Institution, Major R. St. G. Barthelot was of great help subsequently in answering questions and providing me with books and illustrations.

My late friend, Major G. Tylden, E.D. helped me much with the loan of books and from his own encyclopaedic knowledge of military affairs. Finally I must mention the following, to which I have the honour to belong: The Royal United Services Institute for Defence Studies, The Society for Army Historical Research, the Military History Society of Ireland, the Navy Records Society, and the London Library.

CHAPTER I

The First Ordnance

THE first certain record of ordnance is contained in some MSS belonging to the City of Ghent and compiled in 1313. An entry in this states: "Item, in this year the use of bussen was first discovered in Germany by a monk." And in another Ghent MS of 1314 there is mention of guns being manufactured in Ghent and exported to England.[1] Ten years later, in 1324 guns are recorded as having been used at the siege of Metz;[2] and in 1326 brass cannon firing iron balls were being made at Florence for the defence of the commune.[3] The nature of these early pieces is not apparent, but even at this period they seem to have been divided into siege and anti-personnel weapons. The earliest gun in Western Europe of which there is an illustration is so small as to have been nearer, in its military use, to a modern light machine-gun than to a piece of ordnance. The illustration is contained in the Millimete MS in the Library of Christ Church at Oxford. Its date can be established as 1327, because the drawing is at the head of a dedicatory address given by Walter de Millimete to King Edward III on his accession to the throne. The little gun is shaped much like a bulbous vase tilted on its side – the bulbous part forming a chamber for the powder and the neck being the barrel. Communicating with the chamber is a touch-hole, to which a man in armour is holding a lighted linstock (a long shaft with a piece of match or other combustible at the end of it). The missile, which is shown as just leaving the barrel, is a heavy feathered bolt, or quarrel, and this establishes the character of the weapon. The French term for such guns was *pots de fer* and the Italians called them *vasi*, both names being derived from the shape.[4] They were obviously developed from the crossbow and the ballista, and it seems doubtful whether they were any improvement on either. They cannot have been very mobile, for the gun in the illustration seems to have been temporarily secured to a wooden trestle. Some of these primitive pieces may have been taken by Edward III on his invasion of Scotland in the first year of his reign, and this would lend point to the choice of this new weapon of war to adorn the address. John Barbour, Archdeacon of Aberdeen, says in his *Metrical Life of King Robert Bruce* that the English used 'crakys of war' in the campaign, and this may or may not be a contemporary Scottish term for guns.

The weapons referred to above as used at Metz and Florence were certainly of a different type – a large clumsy squat mortar which fired iron balls. The brass pieces made at Florence were probably the first made of this metal.[5] Others were made of a form of toughened brass (resembling the later bronze or gun metal), or of latten, which was very like brass.[6] All these were of cast metal but others were of a more complicated wrought iron construction, with strips fitted together lengthwise and hooped with iron rings.[7] Spanish evidence shows that the Moors also possessed pieces which fired iron balls.[8] These mortars were siege weapons, designed for firing heavy stones or iron balls against fortifications.

Although brass and similar metals could be cast, the art of casting iron had not yet been acquired and it could only be used in its malleable or wrought condition. Cast iron is normally obtained by melting ores in blast furnaces and then running the metal into rough bars or 'pigs'. During melting various fluxes are added to carry off a portion of the carbon and other impurities. The next stage is to pour it into moulds; but in early days it was not practicable to heat the iron to a sufficiently high temperature for this purpose. It was many years before this difficulty was overcome. To get usable iron, therefore, still more carbon had to be removed until less than 2 per cent remained. The iron could then be worked by hammering or rolling it; hence the term 'wrought'. It also had the property of welding. Even in primitive times it was quite simple to obtain iron from the ore in small quantities. With the use of a quantity of wood charcoal and the industrious application of his bellows, the smith could make his furnace hot enough to extract iron of the purest quality. This he wrought on an anvil and drew into a bar or plate as required.

From the squat weapon, which looked like a chemist's mortar and was so named, there was soon developed a longer tube-shaped gun, made in a somewhat similar fashion to that used for the mortars. Wrought iron bars were fitted next to one another, when red-hot, round a wooden mandril, and the gunsmith then beat them into a solid tube with his hammer. Round the tube were then placed white-hot hoops of iron which were allowed to shrink on. From the resemblance of this form of construction to that used in the manufacture of wine barrels, the name 'barrel' was given to the tube. In earliest days the hoops numbered between two and six, but later they were fitted close to one another so that the whole barrel appeared as a rigid hollow cylinder. After the hoops had cooled the mandril was bored out. Since this tube, or barrel, was open at both ends, a separate chamber was made to hold the powder. It was forged out of solid metal and tapered to fit the rear end of the barrel, or breech. Most of these early wrought iron guns were therefore breech-loaders. The breech was held in position by a wedge driven between it and an upright at the rear end of a wooden gun cradle, to which the gun was secured by a hempen rope.

By the end of the fourteenth century some of these wrought iron guns had reached a considerable size. Ralph de Halton, Keeper of the Privy Wardrobe to King Richard II, bought seventy-three guns between 1382 and 1388 from a gun-founder named William Woodward, and of these, forty-seven averaged 380 lb each.[9]

There are in the Rotunda at Woolwich two wrought iron 'serpent' guns of the reign of King Henry VI (1422–61) with a calibre of 4.25 in., lengths of 7 ft 6 in. and 8 ft 6 in., and weights of 8 cwt and 8 cwt 70 lb respectively. The construction of the former of the two guns was examined in 1861; this had had 28 inches broken off at the breech end a long time previously, so exposing the manner of construction. It was found that there were fourteen longitudinal bars arranged in a circle, two deep, like the staves of a barrel, and that these had been imperfectly welded together, leaving interstices into which molten lead had been poured. Over the internal tube, so formed, were 35 hoops, averaging 2.3 in. wide and 1.5 in. thick at the breech end. These hoops had been burred down at the edges while hot and brought together to a very close union. A bronze cylinder was inserted at the breech end to serve as a powder chamber.

Brigadier O. F. G. Hogg, in *English Artillery 1326–1716*, says that these early wrought iron guns were known as 'bombos' – doubtless an alliterative term, meaning a loud humming noise. Major-General J. F. C. Fuller, on the other hand, refers in his *Decisive Battles of the Western World* to 'lombards', or 'bombards'. Certainly the Lombards played a prominent part in the development of these early pieces.

The gunpowder used for these guns consisted of almost equal proportions of saltpetre, sulphur, and charcoal, and was ground into a fine powder. It was dangerous stuff; so apt to explode, once mixed, that the charcoal was sometimes kept separate from the other two ingredients until required. There was another reason for not mixing it until just before use: when it was jolted about in transport its three ingredients tended to separate into three separate strata, with the charcoal on top, the saltpetre in the middle, and the sulphur at the bottom.

Ramming the gunpowder into the gun had to be done with some care. If this fine stuff was packed too tight it tended to lose its explosive character, because the flame which ignited the powder from the vent could not spread quickly enough through the whole mass. If, on the other hand, the packing was too loose, the flame might peter out before it had ignited the bulk of the powder; again there would be no effective explosion.[10]

The powder charge was probably ignited by applying a hot iron to a train formed by grains of gunpowder leading through the touch hole to the charge. But slow match was introduced at an early date, and this was ignited by another piece of slow match held in a linstock. Slow match consisted of cotton soaked in saltpetre and dipped in melted sulphur. Early cannon balls were of iron, but they were expensive and by about 1350 stone balls were commonly used.[11]

The method of mounting these early guns was extremely rudimentary. There were no such things as gun carriages, and that simple but invaluable addition to the barrel, the trunnion, had not as yet been invented; it did not in fact appear until the middle of the fifteenth century.[12] Guns were secured to a primitive wooden cradle, which was variously a simple beam (known first as either a trunk or tiller, and later a gunstock), a sledge, or a stand. In fixed defences the gun might merely be laid flat on the wide embrasures of the gun port.[13] Sometimes the gun was merely placed on the ground, lashed to wooden frames known as 'telaria' which were erected on each side of it; the breech end being butted against a wooden support.

The trunk or tiller was eventually a wooden baulk hollowed out and bound with iron, and to this the gun was secured either by hempen ropes or iron straps. To lay the piece the muzzle end was raised on crossed stakes, while the breech end was anchored.[14]

By the end of the fourteenth century guns had already acquired the peculiar names of birds and reptiles by which they were to be known until supplanted by a nomenclature based on the weight of the shot and the calibre of the bore. These early names are confusing because they did not always refer to pieces of the same size or bore, and their usage changed in some cases as the years went by; some new names appeared, and some of the older ones were discontinued. An early list is as follows: Syren (60-pr), Basilisk (48-pr), Cannon Royal or Carthoun (48-pr), Bastard Cannon or ¾ Carthoun (36-pr), Half-Carthoun (24-pr), Whole Culverin (18-pr), Demi Culverin (9-pr), Saker (large 8-pr, smaller 6-pr, ordinary 5-pr), Dragon (6-pr), Serpentine (4-pr), Falconet (3-pr, 2-pr, and 1-pr),

Aspic(2-pr), Robinet(1-pr), and Moyen(12-oz).[15] Gradually
'cannon' and its varieties came to be applied to short pieces,
whereas 'culverin' and its varieties denoted long pieces.

In 1339 there is the first mention of a peculiar weapon
called a ribauld or ribauldequin, which was an anti-personnel
weapon and might perhaps be regarded rather as the
ancestor of the machine-gun than as ordnance. It consisted
of from two to several small pieces cradled in a group so
that the gunner, with one sweep across the vents with his
linstock, could fire them almost simultaneously. The platform on which they were mounted was furnished with
wheels and so formed a primitive gun carriage, and it
was often protected by a mantling which gave cover to
the gunners – the first example of a gun shield. This first
mention of ribaulds is recorded in the city accounts of
Bruges as purchased for the defence of the town. In 1340
they were used during the siege of Tournai in a work
placed to command the gates of the city.[16]

In 1339 also the French were using small guns in their
attack against the English fortress of Puy Guillem in
Périgord, and the Scots in their siege of Stirling.[17] Edward
III collected a train of artillery in 1345 for his invasion of
France. Robert de Mildenhall, Keeper of the Tower
Wardrobe, was directed to construct 100 ribaulds, collect-
ing for the purpose whatever timber, wheels, axles, and
other materials were necessary. They appear to have been
used at the siege of Calais, particularly in the lines of
contravallation which protected the besieging troops from
interference by a relieving force.[18]

Apart from the ribaulds, Edward had a considerable
number of ordinary guns at the siege of Calais, and there
is some evidence that a few were in action at the Battle of
Crécy. Froissart mentions them in one account of the battle
but not in a later version. The *Grandes Croniques de France*,
regarded by Oman as 'a pretty good authority', says that
at the beginning of the battle the English '*getterent trois
canons*'. Supporting evidence is provided by the digging
up on the battlefield of small iron and stone cannon balls
about 1½ lb in weight. When Edward laid siege to Calais

he sent to England for guns and ammunition and it is recorded that bombards were sited at the sea ends of the siege works to command the entrance to the port of Calais.[19]

After this campaign waged by Edward III, there are numerous mentions of artillery in European wars. In 1374 the French used large bombards in their attack on Sir John Chandos's Castle of Saint Sauveur le Vicomte. But the heavy stone balls which were fired were inclined to break up on impact, and the castle withstood this battering successfully for over a year. In 1372 an English expedition was supplied with 29 guns, and in 1375 two 'great' and two 'less' cannon were sent to Brest with an ammunition supply of 600 stones for these and 'other engines'.[20] In Scotland 'small cannons' were used during the reign of Robert II between 1370 and 1390.[21]

The effect of the new weapon was strikingly demonstrated in England early in the fifteenth century. In 1405 the siege cannon of King Henry IV rapidly reduced the Duke of Northumberland's formidable castles at Berwick, Alnwick, and Warkworth. The garrison commander at Warkworth Castle had expressed his confidence in being able to defend it successfully, but the destruction brought by only seven of the great projectiles convinced him of the uselessness of resistance and he surrendered.[22]

In the next year Henry IV, now on the throne, had a train of artillery loaded on ships at Bristol for his siege of Aberystwyth.[23] In 1415 a siege train under the command of his brother Humphrey accompanied the army to France for the siege of Harfleur. The guns played a major part in the capture of the town, the gunners serving them being protected by shelters and working in relays to maintain a continuous bombardment.[24] These guns, however, were only for siege operations and did not accompany the field army. The French had a few guns at the battle of Agincourt, but they were quite ineffective.[25]

When the English evacuated Mont St Michel in 1424 they left behind them two bombards (which still exist), the larger of which is 12 ft long, has a calibre of 18 in.

and weighs 5½ tons. They are possibly of English make and date perhaps from the late fourteenth century. The larger piece is of the built-up wrought iron type. The interior is constructed of longitudinal bars 2¾ in. wide and 1 in. thick, and round these are iron hoops fitted close together. The gun is, of course, a breech-loader, with a 3 ft long chamber made of longitudinal wrought iron bars. There are as yet no trunnions, but on either side of the gun is an eye to take the tackle for securing it to its mounting. The chamber was held in position by a wedge driven between it and a vertical support in the rear.[26]

CHAPTER II

Artillery Comes of Age

By the time that King Henry VI came to the throne artillery was an essential arm in warfare; but it was mainly a siege weapon and had not as yet established itself as an important weapon of the battlefield. The main reason for this was probably the absence of an effective gun carriage. Towards the end of the fifteenth century, however, wheeled carriages were coming into widespread use. These somewhat primitive vehicles consisted of a solid baulk of timber, which formed the trail, an axletree, and a pair of heavy and clumsy wheels. The gun rested in a wooden cradle, which itself had trunnions, and was borne in bearings in the trail over the axletree. Even by the first quarter of the fifteenth century such a simple device as a trunnion had not been invented for guns, and they were commonly secured to their mountings by eyes, like the Mont St Michel bombards.

Relics still extant show the trunnion as it was in its early days in the Burgundian artillery of Charles the Bold (1433–77). Trunnions, as stated above, were first fixed to the carriage; and with these, and a curved rack on the trail to support its rear end, the gun could be elevated. Later the trunnions were cast on the gun itself, and the cascabel (rear end of the gun) was supported by a cross-pin between the flanks of the trail. This cross-pin could be fitted into any one pair of a series of holes to alter the elevation. The first trunnions were cast level with the gun axis (as in a gun of Charles the Bold captured in 1476 and in one belonging to Louis XI cast in 1478), but in later guns they were level with the bottom of the bore of the piece. The reason for this change was to ensure a

19

downward pressure on the trail when the gun was discharged. It certainly achieved that object; indeed, when the force of gunpowder was increased, the pressure became so great that the carriage was often broken.[1]

In the middle of the fifteenth century a great improvement was effected by the introduction of corned powder in replacement of the fine, or serpentine, powder. This new type was made by damping fine powder and working it into grains, which were then crushed to a suitable size and sieved. The grains were then glazed to prevent deterioration from damp. Initially the new powder was restricted to the smaller guns; it was some time before the large pieces could be cast in sufficient strength to withstand its force.[2] Thus two types of ordnance were developed at the same time: large stone-throwing pieces of cast bronze or wrought iron, charged with serpentine powder, and small portable or semi-mobile guns using iron shot with corned powder. This dual development continued until the middle of the sixteenth century.[3]

The bronze used for guns was frequently called 'gun metal' and consisted of about 90 per cent copper and 10 per cent tin. It is a tough and tenacious metal, but when cast in the normal fashion it is comparatively soft and was thus liable to be damaged by the projectile. When heated by rapid firing it became still softer. Tin, having a much lower melting point than copper, sweated out in parts when the gun was cooling, causing white spots known as 'tin spots'. These were acted upon by the gas of the powder and gradually eaten away, leaving flaws or even holes. For small-bore field guns, however, having small shot and therefore small charges, bronze answered well enough.

In Scotland in 1430, James I imported from the Continent a 'monstrous brass gun' which was given the name of 'The Lion', though sometimes referred to in contemporary chronicles as 'The King's great bombard'. It was clearly thought to be of great size; judging by contemporary foreign guns, it was probably a breech-loader firing a 500-lb shot.[4] From 1444, bronze (commonly called 'brass') was used more frequently on the Continent; owing primarily to the work of Jean Bineau.[5]

The manufacture of ordnance in Scotland is first expressly recorded in 1471, when a considerable number of the small ribaulds, or 'cart' guns, were made. But their manufacture may have started earlier, for in 1456 the Scots Parliament thought it expedient for the King to request certain of the great Barons to make 'carts of war'. This was conveyed in an Act of 19 October, 1456 which specified that in each cart there should be two guns, each having two chambers, and that there was to be a knowledgeable man to shoot them.[6]

To James IV, who ascended the throne of Scotland in 1488, was due the construction of guns in Scotland on a considerable scale, most of them being made of bronze.[7] These guns were cast and were muzzle-loaders.

The Lombards, in the fifteenth century, cast bombards, in bronze, with most elaborate mouldings. In 1430 a famous breech-loading bombard was made by Flemish artisans at Ghent and named 'Dulle Griete'. It fired an enormous stone shot 25 inches in diameter. The chamber of this gun screwed on to the barrel, the screw threads being formed on a boss at the front of the chamber and in a hole at the rear of the barrel.[8]

The celebrated 'Mons Meg' at Edinburgh Castle is a built-up wrought iron bombard, believed to have been made at Mons, then a big foundry centre, in about 1460. It is first mentioned in connection with the siege of Dumbarton by James IV in 1489. It is constructed of longitudinal wrought iron bars 2½ in. thick, which have been welded together over a mandril, and over which have been shrunk wrought iron hoops each 3½ in. thick. The gun is 13 ft 2 in. long, has a calibre of 19½ in., weighs 5 tons, and fired a granite ball of 330 lbs. 'Mons Meg' was unique amongst Scottish wrought iron bombards in being a muzzle-loader. The accident that killed James II may have been responsible for its purchase. The King was standing beside one of his bombards at the siege of Roxburgh in 1460 when the wedge securing the breech was blown out and his thigh

21

A breach made by mortars in a city wall.

bone was 'dung in two', inflicting a mortal wound.[9] 'Mons Meg' was last fired in 1680 in honour of the Duke of York, and it was on this occasion that the existing damage in its side was caused. The present carriage is not the original, but was made at Woolwich in 1836.[10]

At about the end of the fifteenth century an improved type of construction for wrought iron guns was introduced. Instead of iron bars a number of very short lengths of tube with lips at their extremities were fitted together and held in position by broad hoops shrunk over adjoining lips.[11]

Built-up wrought iron guns were made as late as the sixteenth century; but by then this type of construction was well out of fashion. The great weakness of wrought iron guns was that they were almost invariably breech-loaders. As already pointed out, it was difficult to build a wrought iron gun except as a tube open at both ends, with a separate chamber; and the manufacture of a satisfactory chamber, proof against a large leakage of gas from the combustion of the powder, was beyond the art of the gunmakers of the time. Apart from their inefficiency, these clumsy-looking wrought iron bombards were quite outclassed in appearance by the beautiful cast bronze muzzle-loading guns of Renaissance workmanship which were being made by the end of the fifteenth century.[12]

In the French military renaissance, inspired by St Joan of Arc, Jean and Jasper Bureau were responsible for a vastly improved artillery which comprised, not only huge bombards for siege work, but for the first time guns which could be, and were, used with effect on the battle-field.[13] In their reconquest of Normandy in 1449–50 the French conducted sixty successful sieges. The great siege guns were, in fact, so feared that many fortresses surrendered as soon as the batteries were in position.[14]

The Battle of Formigny on 15 April, 1450 deserves a special mention in artillery history. It was one of the last battles waged by the English to hold their territories in France. An English army about 4,500 strong had been assembled to try and relieve a force at Caen, under the Duke of Somerset, which was besieged by an overwhelmingly superior army under the French King. Near the village of Formigny the relieving English troops were confronted by the French under Clermont in slightly superior strength and with two field-guns. The English fell back to a defensive position in their well-tried formation of dismounted men-at-arms flanked by wings of archers. The French, however, did not deliver one of their usual frontal attacks; instead, after some lengthy skirmishing Giraud, Master of the Royal Ordnance, brought up his two culverins to a position from whence they enfiladed the English line. Suffering heavy casualties from this fire, some of the archers charged and captured the guns. But the culverins had achieved Giraud's object, for the French counterattacked the broken line and inflicted an overwhelming defeat on the English force.[15]

23

At about this time canister (or case) and grape shot were introduced to supplement the solid iron and stone shot. Canister, as its name implies, was a container filled with all sorts of scrap, such as nails, bits of iron and even gravel. Before the introduction of a canister this collection of junk had been loaded loose into the barrel of a gun, and in this form was known as 'langridge'. Grape shot was somewhat similar to canister, but consisted of a far more sophisticated arrangement of round shot packed in nets or sacks, and was so-called from its fancied resemblance to bunches of grapes.[16]

In 1494 Charles VIII of France invaded Italy with an army which included both field and siege artillery. The artillery pieces consisted of stone-throwing mortars for siege work and 8 ft-long, mostly bronze, guns mounted on carriages and drawn by horses for use in the field.[17] This train, which, unusually for the time, could keep up with marching infantry, impressed the Italians. In their own campaigns the artillery in the *condottieri* armies had been confined to a few big guns drawn by oxen and pulled slowly along well behind the main body of the army. When it eventually arrived on the battlefield it took some time to get it into action, and Machiavelli, in his *Arte della Guerra*, said that it could only get off one or two discharges before battle was joined. The French guns not only moved much more quickly but the gunners had been trained in rapid loading and traversing, and had reached a degree of proficiency previously unknown.[18]

The Battle of Fornova, on 8 July, 1495, between Charles and the Italians, may have been the first instance of the really effective use of artillery in the field. The *Mer des Histoires* (published in Lyons without date) says that 'a square of pikes, with Swiss culveriniers on the wings, held the Venetian cavalry in check. Four pieces flanked one side of the square and two others were "en batterie" on the left wing; for the artillery was in action from the start of the battle'. An ancient engraving of the Battle of Fornova shows these six pieces in action on wheeled carriages. They have no trails but are supported instead by two struts in front

Cannon at the time of Fornova. From an old engraving.

resting on the ground on either side of the barrel. (Perhaps these struts were lengthened to form shafts for moving the guns.) This is the earliest known illustration which shows a battle with field guns in action.

In England, during the Wars of the Roses, artillery, both siege and field, was little used. At the Battle of Northampton in 1460 the Lancastrians built strong lines of defence in which big guns were sited, but nearly all the powder was damaged by rain and few shots were fired.[19] Moreover, there were few sieges because of the general conviction (resulting from the French wars) that the old fortifications and walls were useless against big guns. Bamburgh Castle did attempt to hold out for the Lancastrian cause in the Northumbrian campaign of 1464, but Edward iv's siege guns soon destroyed it.[20]

That the French field train was somewhat unique is shown by the use the Spaniards made of artillery in the conquest of Granada in 1492. The commander of the Spanish artillery was Don Francisco Ramirez, and his train for the campaign was 'such as was probably not possessed at that time by any other European potentate'.[21] The largest bombards had a calibre of 14 in., were 12 ft long, and were of the wrought iron construction of the period. They were mounted on wooden carriages without wheels and could be neither elevated nor traversed. They were dragged laboriously by vast teams of oxen, and special roads had to be built for them by thousands of pioneers. But without them Ferdinand's reduction of the hitherto impregnable Moorish castles would have been impossible.[22]

CHAPTER III

Tudor Ordnance

O^N 11 April 1512 artillery became established as a decisive weapon of the battlefield, for on that day was fought the Battle of Ravenna at which the Spanish defeat was due primarily to the artillery of the French and their allies. It was probably, indeed, the first general action in history of which the outcome was decided by ordnance.

The French army under Gaston de Foix, Duke of Nemours, was engaged in besieging the fortress of Ravenna, when they learned that a Spanish army was advancing to its relief. Gaston decided to attack the Spaniards rather than await their arrival. On his approach, however, the Spanish army took up a strong defensive position in front of a river with the flanks resting on it. Gaston decided to open an overwhelming fire at the weaker points of the Spanish position with his artillery and then deliver a frontal attack with his infantry and cavalry. (It was a tactic that was to become only too familiar some four hundred years later.) He had a very strong artillery train which comprised thirty of his own guns and twenty-four of the celebrated train of his ally, the Duke of Ferrara.

The French bombarded the Spaniards for two hours, and in additon to the frontal fire, Gaston sent two guns across a bridge to enfilade their position.[1] The *History of Bayard*[2] says that these were 'a cannon and a long culverin'. Of the battle it says, 'The Spanish had already begun to draw from their camp . . . . Behind them were all their footmen, lying upon their bellies for fear of the artillery of the French. In front of them was all their artillery, in number twenty pieces, as well cannon as long culverins'. The French gained a decisive victory, though – another pattern of the distant future – their infantry incurred heavy casualties.

In the construction of ordnance the English appear to have been behind the Scots. Grose, in his *Military Antiquities*, says, 'Although artillery was used from the time of King Edward III and purchased from abroad by all our successive kings, it seems extremely strange, that none of our workmen attempted to cast them, till the reign of King Henry VIII, when in 1521, according to Stowe, or 1535 (Camden says), great brass ordnance, as canon and culverins, were first cast in England, by one John Owen.' The first English cast iron guns were also made in Henry VIII's reign. In the first half of the sixteenth century the Weald of Sussex was the most important centre of iron smelting in England. Deposits of iron ore and the heavily wooded countryside provided both the raw material for the metal and the fuel with which to process it. The Weald, indeed, remained a centre of metal work until in the seventeenth century coal replaced wood as a fuel.[3] Here were made the first English cast iron ordnance. Cast iron, apart from being cheaper, was harder and more enduring than bronze; but it had the disadvantage of being more likely to crack and fly apart. Also it needed an enormous amount of wood charcoal to produce it, until the superiority of coal was discovered. These first English cast iron guns were made by Ralph Hogge at Buxted in Sussex in 1543.[4] The industry flourished in the Weald for a long time and both iron and bronze guns were still being made there in 1614.[5]

But there were still very few gun-founders in England in the early years of the reign of King Henry VIII, and he bought most of his ordnance from the Low Countries. In 1512 he purchased from Hans Poppenruyter of Malines 16 guns (which he named 'Wales', 'Cornwall', 'Chester', 'Sun', 'Portcullis', 'Garter', 'Ireland', 'Rose', 'Crown', 'York', 'Lancaster', 'March', 'Guienne', 'Richmond', 'Normandy', and 'Clarenceux') all of which weighed between 3,000 and 4,000 lb, and 12 guns named after and known as the 'Twelve Apostles'. The Netherlands even supplied the horses, for, also in 1512, the Exchequer Accounts record the purchase of 14 Flemish draught mares for each gun.[6] This would provide one horse for about every 250 lb of gun (less carriage).

Henry VIII took a great interest in these guns. In 1518 he travelled to Southampton to watch them firing and noted the ranges obtained. His gunners were mostly foreigners who had had practice with guns of this type.[7] The King's interest led him to persuade a number of foreign gun-founders to come to England and start building up a gun-founding industry.[8]

The Flodden campaign took place in 1513, and guns of England and Scotland were ranged against one another. The artillery in the Scottish army was probably the better, owing to the attention which James IV had given to this arm of the service over so many years. Not only were the guns good, but the Scottish artillery was the only arm in either country (except for Royal Bodyguards) which was composed of regular troops.

The overwhelming defeat of the Scottish army resulted in all the 17 field guns which had accompanied it falling into the hands of the English. These were all noted in the Lord Treasurer's list. The largest were 'courtaulds' or 'murtherers', weighing about 6,000 lb with a calibre of 6½ in. and firing a projectile weighing 33½–36 lb. A smaller gun was listed as a 'gros culverin of found' (i.e., a large cast culverin). As it required as many oxen to draw it as the courtauld it must have been about the same weight and, as a culverin, of smaller calibre and longer barrel. Still smaller was the 'culverin pykmoyane', also known as a 'saker'. It weighed about 2,850 lb and fired a 7 or 10-lb ball. The lightest of these field pieces was the 'culverin moyane', of cast bronze, weighing 1,500 lb, having a calibre of 2½ in, and firing a projectile of about 5 lb. In addition to these 'conventional' pieces, there were the little 'cart' guns, alternatively referred to as 'falcons', and which were breech-loaders.[9]

The type of gun carriage used during the Flodden campaign presents something of a problem. The old bombards were generally transported on carts or wagons and hoisted on to a stationary mounting when brought into action. It would appear that the biggest guns were still moved like this during the Flodden campaign, for just before it the Accounts of June 1513 record the transport of two 'great guns' in carts from Threave Castle to join the train of artillery for the field army; and the following month a cannon was carried in six carts drawn by thirty-six horses from Edinburgh to Glasgow on its way to join a force assembling for Ireland. The gun was accompanied by a crane for mounting and dismounting the gun and the 'traistes' (i.e., trestles) on which it was mounted.[10] It seems reasonable to suppose that the smaller guns, at least, were mounted on wheeled carriages, for it is recorded that the guns came repeatedly into action during the Battle of Flodden and, owing to the 180° change of front by the army, static mountings could hardly have been moved in time.

Ox-yokes for gun teams were under construction in Scotland in 1512, and the following scale was laid down for the animals to be used in the haulage of guns:

Cannon and great culverin	36 oxen
Culverin pykmoyane	16 oxen, 1 horse
Culverin moyane	8 oxen, 1 horse.

This supports the supposition that the lighter guns were mounted on carriages because the single horse was presumably placed between the shafts. The heavy guns, with no horse in the team, were probably carried on wagons and transferred to fixed mounts on the battlefield.

The cannon sent to Glasgow for shipment was accompanied by eight 'close' carts, each carrying one barrel of powder, and two carts laden with 'gun stones' weighing 33½ lb apiece. For the Flodden guns there were thirteen carts, each carrying four barrels of powder, and twenty-eight pack horses carrying projectiles in creels, or panniers.[11]

Albrecht Dürer (1471–1528), that prolific artist who portrayed exactly what he saw and in great detail, drew a picture of the artillery in the army of the Emperor Maximilian (1459–1519). He shows cannon with lifting handles, or eyes, on either side of the top of the barrel, and one of these is suspended from a tripod crane. Other guns are on static

The Emperor Maximilian with his artillery. After Albrecht Dürer.

mounts, and there is also a mortar mounted and fitted with trunnions. The 'Loyal Serviteur', in his *History of Bayard* (Pierre du Terrail, Chevalier de Bayard, 1473–1524) of 1527 says,

> He [Maximilian] had one hundred and five pieces of artillery on trucks, the smallest of which was termed a falcon, and six large bom-ketches, which could not be fired from off the gun-carriages, but were drawn on powerful waggons and accompanied by cranes, and when they wished to fire them they were placed on the ground, and with the crane was raised slightly the mouth of the piece, under which was placed a large piece of wood, and behind them placed a strong buttress to prevent recoil. These pieces fired off bullets of stone, of weight hardly to be lifted, and could only be fired at most four times each day. . . . The Cardinal de Ferarra came in place of the Duke, his brother, to the Emperor's assistance, bringing with him twelve pieces of artillery, five hundred horses, and three thousand foot-men. . . . A great defect was as regarded the artillery, for they had but waggons for half the cannons, and, when marching, were obliged to leave part of their forces to protect the ones left till such time as the first half were placed on the

field there to remain. When the waggons returned to fetch the others, this caused much vexatious delay.

The particular interest of this account is that it implies that the smaller guns were fired from carriages and the larger ones, in primitive fashion, from the ground.

Contemporary bas reliefs of the Battle of Marignano, which took place in 1515, two years after Flodden, between the French army of Francis I and the Swiss mercenaries of the Duke of Milan, show cannon in action on wheeled carriages with trails, and having powder casks beside them. The attack of the Swiss pikemen was repulsed by fire from these guns, supported by cavalry charges. The wheels obscure the sides of the gun barrels, so that it cannot be seen whether they have trunnions or not, but there is no other visible means of attaching them to the gun carriages. The reliefs also show artillery on the march, the guns having, apparently, shafts attached directly to the trail, with one horse inside them and one outside them on the near side, the pair forming the wheelers of the team. The evidence of these reliefs is supported by a sixteenth-century miniature in the Musée Condé at Chantilly which shows similar guns in action on each side at Marignano.

There was a new development in England when Henry VIII appointed two German specialists, Peter Bawd and Peter van Collen, to the Royal Foundry and Arsenal at Greenwich to make large mortars and shells. They are reputed to have made in 1543 the first explosive shells for mortars. Accord-

A bas-relief showing cannon at the Battle of Marignano.

32

ing to John Stow (1525–1605) these were 'hollow shot of cast iron stuffed with fireworks, fitted with screws of iron to receive a match to carry fire kindled, that the fireworks might be set on fire to break in pieces the small hollow shot where of the smallest piece hitting any man did kill or spoil him'. The method of igniting the fuse was either to place the shell in the bore with the fuse towards the charge, or, more dangerously, to put it with the fuse towards the muzzle and light it by thrusting a match down the bore.

The introduction of cast iron guns was very slow, and the guns which were actually being used in the middle of the sixteenth century can be deduced from the pieces of ordnance recovered from the ship *Mary Rose*, sunk in action with the French fleet off Spithead in 1545. It is apparent that cast iron guns must have been still very rare in the English service. Some of the *Mary Rose*'s guns were the old built-up wrought iron breech-loaders. Of these, most were of bar-construction with a calibre of 8 in. and a length of 9 ft 8 in.; but at least one (now at Greenwich) was the short-tube type, 7 ft 6 in. long and of 5-in. bore. But 'brass' (i.e., cast bronze) muzzle-loaders were by far the most numerous, and these were of the following categories: culverin, calibre 5.2 in., weight 4,800 lb, length 16 ft 11 in.; demi-cannon, calibre 6.4 in., length 11 ft; culverin bastard, calibre 4.56 in., length 8 ft 6 in.; and cannon royal, calibre 8.54 in., length 8 ft 6 in. All of these bronze guns were very ornamentally emblazoned.[12] These calibres and lengths are of interest in showing the difference between the short large-bore cannon type and the long small-bore culverin, with the 'bastard' pattern half-way between the two.

In the mid-sixteenth century ordnance could be divided into four major classes: cannons, culverins, perriers, and mortars. The cannons had comparatively short range but heavy hitting power, the culverins threw a smaller shot at a longer range, the perriers (i.e., stone-throwers) were the predecessors of howitzers, and the mortars fired projectiles at a very high and fixed trajectory. Pieces used in warships of 1559 included a 4,500 lb 17⅓-pr culverin with a range of 2,500 paces, a 4,000-lb 30⅓-lb demi-cannon with a range of 1,700 paces, a 3,400-lb 9⅓-pr demi-culverin with a range of 2,500 paces, and a 3,000-lb 24¼-pr cannon-perrier with a range of 1,600 paces.[13] None of these were cast iron. Cannons, when carried, were built-up wrought iron, whilst demi-cannons and culverins were cast bronze.[14]

The powder charge used with the guns of the mid-sixteenth century was of roughly the same weight as that of the shot. The 'windage' (i.e., the air gap between the round loose-fitting projectile and the barrel) of a gun caused so much wastage of explosive effort that a very large charge was necessary.[15] Until the end of the eighteenth century the windage was fixed at ¼ in. between shot and gun, irrespective of the size of the gun. The smaller the gun, therefore, the bigger the proportion of windage to bore, and so the larger the relative charge. An Apostle fired a 20-lb projectile with 20 lb of powder; a curtow, one of 60-lb with 40 lb of powder; a culverin, a 20-lb shot with 22 lb. The number of times per day that the larger guns could safely be fired was laid down: e.g., an Apostle, 30; a curtow, 40; a culverin, 36.[16]

The Battle of Pinkie between the English and the Scots on 10 September 1547 was probably the first in which English artillery played a decisive part. The Scots took up a position on the line of the River Esk, which falls into the sea between Musselburgh and Inveresk. The Duke of Somerset, commanding the English force, decided to deliver a frontal attack in a Ravenna-type battle. Unexpectedly, however, the Scots took the offensive, with the result that, both sides advancing, there was an untidy encounter battle. The English warships in the bay fired with effect at the Scottish left as it crossed the river; while Somerset tried to halt the Scots by a succession of cavalry charges, supported by artillery fire at their flanks and at such parts of their front as were not being attacked by the charging squadrons. These tactics were successful, and the Scottish army was defeated primarily by the English artillery fire from land and sea.[17]

5. A sixteenth century cannoneer laying his piece. From a contemporary drawing.

Guns were first systematically classified by the Emperor Charles v. He was also probably the first to use artillery as an efficient weapon on the battlefield. His field batteries consisted of bronze guns with trunnions, mounted on wheeled carriages and firing cast iron balls. Having experienced the inconvenience of having many different types of gun with a multiplicity of calibres, he limited his ordnance in about 1544 to seven standard pieces: a 40-pr cannon, a 24-pr cannon-moyane, two types of 12-pr culverins (with probably different lengths of barrel), two types of 6-pr culverins (differing no doubt as the 12-pr), and a 3-pr falcon.[18]

This idea of standardization was taken up by the French. In 1550 King Henry ii approved the following six types of ordnance as the only ones to be used by the French army:

Cannon – 33-pr, $10\frac{1}{2}$ ft long, weighing 5,200 lb, drawn by 21 horses.
Culverin – 15-pr, 11 ft long, weighing 4,000 lb, drawn by 17 horses.
Bastard culverin – 7-pr, 11 ft long, weighing 2,500 lb, drawn by 11 horses.

Culverin moyane – 2-pr, 8½ ft long, weighing 1,200 lb, drawn by 4 horses.

Falcon – 1-pr, 7½ ft long, weighing 700 lb, drawn by 3 horses.

Falconet – ¾-pr, 7 ft long, weighing 410 lb, and drawn by 2 horses.

All of these were mounted on wheeled carriages from which they could be fired, so that a French army could be accompanied by a formidable field train. Movement, however, was rather primitive, for there were no limbers and the guns were pulled, muzzles foremost, with their heavy trails dragging along the ground.[19]

The method of hauling guns at this period seems to have varied considerably. Some evidence can be gleaned from *Louis Napoléon sur l'Artillerie – Atlas*, from which ten plates have recently been reproduced with an explanatory text by Mr W. Y. Carman. Of particular interest is a sixteenth-century illustration which shows a team of four horses pulling a gun through an elementary limber consisting of two wheels and an axle tree. (In a later development of the same form of limber, a Prussian gun of 1636 is shown with pole-shaft and swingle-trees for a team of horses.)

In his book, *The Gunner* (1628), Robert Norton says that the gun-founders of *c.*1475–1525 cast ordnance which was much weaker and more slender than in his day, partly because the fine powder with which they were charged was much weaker than the corned powder. This corned variety was double or even three times the strength and about three times as fast in combustion. To withstand it, the metal of the guns had to be doubled or trebled in thickness.[20] But, as recorded in the last chapter, guns were not cast weaker because the powder was weaker, but rather because the gun-founders could not cast big guns strong enough for corned powder. The guns they made were long because the fine powder was slow in burning and needed a long barrel for its complete combustion.

Cast iron was so much cheaper than 'brass' that an attempt was made about the beginning of the reign of Queen Elizabeth 1 to replace the latter altogether by cast iron. But the manufacture of heavy guns in this material still proved too difficult. And so demi-cannons and culverins were still made of 'brass', while the heaviest guns were still made of wrought iron. Cast iron guns were in any case unpopular, because when a barrel failed it blew to pieces without any warning, often causing death or severe injuries. A bronze barrel which was about to fail usually bulged first, so giving sufficient warning.[21] It was not till the eighteenth century that cast iron came into general use, but then it only replaced wrought iron for the heavy guns, whereas the lighter pieces, including all the field artillery, were still made of bronze. The disappearance of wrought iron was accelerated because of the great superiority of corned powder and the consequent wish for pieces strong enough to take it. When wrought iron went, so did breech-loaders; for all cast guns were muzzle-loaders.

Towards the latter part of Elizabeth's reign the various types of ordnance decreased in numbers; demi-cannons for instance, disappeared as being too cumbersome, the smallest guns as being too ineffective. Most new construction was devoted to long-range culverins, demi-culverins and sakers. They were shorter in the barrel than their predecessors, because, owing to the quicker combustion of the corned powder, the projectile did not need such a long travel. It was possible, therefore, to increase the calibre without increasing the weight.[22]

At the time of the Spanish threat to Elizabeth's throne a considerable amount of ordnance was held by the militia, in addition to the guns mounted in coastal fortifications. Plans for coastal defence in 1588 showed some of this essentially field artillery sited to command possible landing places. The militia companies in the Isle of Wight, a particularly vulnerable area, were equipped with some excellent field guns which were kept in special gun chambers in the churches and brought on parade at militia musters. To ensure their mobility in the event of invasion, Sir

George Carey (later second Baron Hunsdon), Captain-General of the Isle of Wight, gave orders in 1583 that captains were to ensure that fields near the coast had gates suitable for the passage of guns, and were to arrange in good time for the carts to carry ammunition.[23] Some of the ordnance in the forts was apparently in poor condition, for Carey complained that he had only four mounted guns, no platforms strong enough to fire them from, and sufficient powder only for one day's action.[24] Such complaints will not seem strange to those who were working feverishly on Britain's coast defences in the autumn of 1940.

CHAPTER IV

Seventeenth Century

FIELD artillery as such first came into use under the masterly hand of Gustavus Adolphus, King of Sweden. Gustavus was the eldest son of King Charles ix of Sweden. He was born in 1594 and succeeded his father on the throne in 1611, at the tender age of seventeen. During his early campaigns he suffered from the slow, unwieldy and cumbrous artillery of the time, and determined to improve its mobility. The first results of this determination were seen in the war against Poland in 1621, when he brought into action some very light pieces known as leather, or *kalter*, guns.[1] These new guns were also used in the Polish campaigns of 1628–9. They were invented by Colonel Warmbrandt, and consisted of a copper tube bound with iron rings and rope and covered with leather. They weighed only 90 lb without the carriage.[2]

Before his German campaigns, Gustavus undertook a much more revolutionary reorganization of his artillery, making it his main tactical arm and the foundation on which he fought his battles. He divided his artillery into three main types: siege, field, and regimental. For the first two he adopted the same three calibres of 24-, 12-, and 6-pr, but the siege pieces were a lot heavier than the field guns.[3] The latter had presumably shorter barrels, to make them lighter and more mobile, and therefore shorter range. The regimental guns were light 4-prs, two in each regiment. They had the revolutionary innovation of fixed ammunition in wooden cases, and were thus able to fire eight rounds for every six shots of a musketeer. These little pieces replaced the famous leather guns.[4] The powder for the ammunition was packed in a flannel bag, which was then

39

A regiment of German artillery. From an old manuscript.

put into the wooden case, followed by the shot. Rapidity of fire was of course vastly increased by this replacement of the tedious business of weighing out a charge of powder and inserting it into the barrel with a long ladle.

In England a decree of 1619 established strict regulation of the manufacture and sale of guns. Gun-founding was to be allowed only in Kent and Sussex; guns were to be shipped only from the Tower Wharf; guns could be sold only at East Smithfield; guns could be proved only in Ratcliffe Fields; and all pieces were to be inscribed with the founder's name, the year, and the weight of the gun. The yard of the Tower gunmen was on the site of the present Liverpool Street Station. Here they carried out their artillery exercise, including the firing of large guns. In the neighbourhood pieces of ordnance were cast and tested, and the legacy of this activity remains in the names of Gun Street, Fort Street, and Artillery Lane – all just east of the station.[5] Later on, guns were proved on other Government grounds in London, and the Tower gunners moved to the Artillery Yard in Moorfields.

In the first half of the seventeenth century another attempt was made to replace 'brass' by cast iron. Some nice cheap guns in the latter metal were actually made, but they could not stand up to corned powder and were rapidly withdrawn.

As regards the movement of guns, Captain Henry Hexham, in his *Principles of the Art Military Practised in the Warres of the United Netherlands* (2nd Edition, printed at Delf in Holland in 1642), says that the block or sling wagon could be used for the carriage of a piece of ordnance or small punts or boats for a river crossing. The piece of ordnance for which this wagon was used was probably a heavy gun which had been removed from its carriage. Hexham describes the block wagon as a four-wheel open-frame vehicle with shafts, and the most useful in a train of ordnance. He says that it could be used also to form blocks against cavalry; and that it was easier to haul a piece 'through moorish, foule, and sandie ways upon this because the peece lieth more steddie, and is not subject to so much

41

An early seventeenth century woodcut showing one way of hauling
cannon.

wrenching aside in durtie and ruttie waies, than upon its
proper carriage. Now whensoever an Enemie should draw
neere unto an Armie by the help of the Fearne you may
quickly hoize it up and laye it upon its own carriage.'
(A fearne was a type of crane used to hoist a piece on and
off its carriage.) Hexham has an illustration of a 'halfe-
Canon' mounted on its carriage and drawn by 'seven
couple of horse and a thiller horse' by means of a small
limber consisting mainly of a pair of wheels and an axletree.
'Thill' or 'thiller' was the old name for shafts, and a
thiller horse was one harnessed between shafts. The

primary purpose of the thiller horse was to act as a brake on the gun carriage by 'sitting on the breeching' (a wide belt fastened below the root of the tail) when going downhill. If the road was bad, a cannon weighing 7,000 lb would need fifteen pairs of horses and a thiller horse, and a demi-cannon eleven pairs and a thiller to pull the 4,500 lb of the gun and 900 lb of the carriage. Other pieces required horses as follows: a field piece, or quarter-cannon, six pairs and a thiller; a falconet, two pairs and a thiller; a small drake of 250 lb, one horse. Artillery played a prominent part in the Civil War in Great Britain of 1642–51, though in siege warfare rather than in field operations. Nevertheless artillery was an essential component of a field army; indeed, owing to the shortage of the bigger field pieces, siege guns were sometimes used on the battlefield. Classification between siege and field guns was, however, somewhat flexible. If a gun was light enough to be hauled along the frightful roads of the time at the pace of marching infantry it could be used in the field. The heaviest piece normally used as a field gun was the culverin, though its appearance in such a role seems to have been rare. It was of 5-in. calibre, fired a 15-lb shot, and had a range of 460 yards point blank and 2,650 yards maximum. It weighed about $1\frac{3}{4}$ tons and was normally pulled by eight horses in tandem. Its maximum firing rate was about ten rounds per hour. The more usual heavy field gun was the demi-culverin with a calibre of $4\frac{1}{2}$ in., firing a 9-lb shot. It weighed about $\frac{1}{4}$ ton less than the culverin and had the rather shorter ranges of 400 yards point blank and 2,400 yards extreme. The light field guns all seem to have been included under the general term of 'drakes', with a maximum rate of firing of about 15 rounds per hour. Their names, weights, and calibres were respectively: saker, 1 ton, $3\frac{1}{2}$ in., 5-pr; minion, $\frac{3}{4}$ ton, 3 in., 4-pr; falcon, $\frac{1}{4}$ ton, $2\frac{3}{4}$ in., 2-pr; falconet, 200 lbs, 2 in., 1-pr; and robinet, 100 lb, $1\frac{1}{4}$ in., $\frac{3}{4}$-pr. The two principal light field guns, the saker and the falcon, had respectively, point blank ranges of 360 and 320 yards, and extreme ranges of 2,170 and 1,920 yards.

The term 'drake' is curious. It originally seems to have denoted merely a lighter gun. For instance, the great *Sovereign of the Seas* of 1637 was armed as follows: on the gun-deck, 20 cannon drakes on the broadsides and 8 demi-cannon drakes as stern and bow chasers; on the middle deck, 24 culverin drakes on the broadsides and 6 culverins as chasers; and on the upper deck, 24 demi-culverin drakes on the broadsides and 4 demi-culverins as chasers. There were 8 demi-culverin drakes on the fo'c'sle, 6 on the half-deck, and 2 on the quarter-deck. The 'drake' versions of these various pieces probably had short barrels, owing to the difficulty of loading long-barrelled muzzle-loaders on board ship.

The siege guns proper were designated cannon, and of these there were three varieties: cannon royal, cannon, and demi-cannon. The respective weights and calibres of these were: $3\frac{1}{2}$ ton, 8 in., 63-pr; $2\frac{1}{2}$ tons, 7 in., 47-pr; $1\frac{3}{4}$ tons, 6 in., 27-pr. The demi-cannon was occasionally used in the field.[6]

Oxen were often used to haul the heavy pieces, and sometimes the lighter ones; but as their pace was slower than that of marching infantry they were less satisfactory than horses in a column of all arms on the road.

A gun crew normally consisted of three men – a gunner, his mate (or matross) and a server. The powder was generally stored in a barrel which was placed behind the gun in action. The gun was charged from this barrel by means of a large iron ladle. Solid iron shot (cannon balls) were normally fired, but canister, or case shot, was also used.[7]

Two small field guns were generally allotted to each infantry regiment; while heavy guns were organized into batteries under the officer commanding the train of artillery. In battle the General of the Ordnance was responsible for siting his guns in consultation with the General of the Army; and in doing so he was largely influenced by the lie of the ground. The larger guns were often sited for overhead fire, provided that suitably high ground could be found, and smaller pieces covered ground where enemy cavalry might break in amongst the infantry. Guns were normally so disposed that some were in front of the infantry line, some between brigades, and some on the flanks or, given high ground, behind. Gun positions were generally 50 – 100 yards apart. All guns had drag ropes attached to them so that they could be easily removed or turned, and each gun position had its escort of firelocks.[8]

William Eldred, in his *The Gunner's Glasse* of 1646, says that the average rate of fire was about eight shots in one hour. But he adds that after firing 40 shots the piece should be cooled for an hour, because otherwise 80 shots might generate sufficient heat to break it.

In the seventeenth century the fuse normally used with the primitive shell of the period was generally cylindrical, about $\frac{1}{2}$ in. in diameter and the length of a finger.[9] As a means of firing the piece, the slow match began to be superseded by the quick match, made by dipping a cotton wick in a solution of saltpetre and spirits of wine and covering it with mealed powder. This was known as 'porte-feu' or 'portfire'. Portfires were made in lengths of 16 in. and were fixed in sticks about two feet long when required for use. They were lit from a piece of slow match (which was always kept burning) when the firing began. At the order to cease firing the burning end of the portfire was cut off with a clipper, which was fixed to the trail of field gun carriages.[10]

A return of British ordnance of 1669 is of interest in showing both that cast iron pieces were in use and that the old names had been abandoned, leaving 'cannon' as a name applied to all guns. The types listed were: brass cannon, 8 in., 7 in., 29-pr, 12-pr, 8-pr, 6-pr, 3-pr; iron cannon, 7 in., 29-pr, 12-pr, 8-pr, 6-pr, 3-pr; brass mortars, $18\frac{1}{2}$ in., $13\frac{1}{4}$ in., 9 in., $8\frac{1}{4}$ in., 6 in., $4\frac{1}{2}$ in., $4\frac{1}{4}$ in.; iron mortars, $12\frac{1}{2}$ in., $4\frac{3}{4}$ in., $4\frac{1}{4}$ in. In 1686 James II ordered the provision of 14 3-pr guns to be provided for attachment to infantry regiments encamped in Hyde Park, on a scale of two to each of seven battalions.[11]

There is a curious note in Pepys's *Diary* of 20 April, 1669 in which he records a visit to 'the Old Artillery Ground near the Spitalfields' to see a new gun 'which from the shortness and bigness they do call Punchinello'. He says that it was tested against a gun which was twice its length and weight and charged with twice the amount of powder. The new gun proved to be the more accurate and was easier to manage. This seems to have been an early version of the carronade of the next century, and it would be interesting to know why it was not adopted.

Captain Thomas Binning, in *A Light to the Art of Gunnery* (1689), has much to say of interest concerning artillery in the second half of the seventeenth century. He gives the following description of artillery on the march: 'The Order and Necessaries for Guns to march by Land, they having six Demi-cannons, six Sakers or Demi-culverings, with two Whole-Cannons, besides the Field-Ordnance.' Before the train marched off it was preceded by companies of pioneers, each man being equipped with a shovel, scoop, pickaxe, crowbar, or handspike. The function of these companies was to make the road suitable for guns. 'After them first follow the 6 Sakers or Demi-culverings with their Provision of Ball in Wagons and their Powder in Wagons'; and behind these came men to remount the guns if they overturned. Next came the six demi-cannons and the two cannons, each followed by wagons containing their ammunition. The end of the column was brought up by 'the Carriage of Ladies' Sponges, and Rammers, Match, Crows, and Handspikes, and Budg-barrels'. On the march every gunner marched 'at the right side of his Peece, and by them their Harbingers, who take notice of all the ropes, and other Provisions for Draught, and help them if defective; and also to see that the Axtrees be well soped or tallow'd'. The Wagon-Master had to have spare horses. Horses for guns were normally calculated on the basis of one horse for every 500 lb of metal. Binning points out, however, that not all horses are the same and that this calculation was based on good roads. He had frequently worked on 350 lb and found the horses barely sufficient for the task. 'If you order a Yoke of Oxen for a Horse-Draught,' he says, 'it will be equal.' Cannons weighed 7,000 lb each, demi-cannons 4,500 lb, and demi-culverins

3,200 lb, or about the same as noted by Henry Hexham, and little different from the weights of 150 years before. If men had to be used instead of animals, Binning reckoned that each man could pull 100 lb on a good road, and he therefore allowed 80 lb for average going.

That the use of limbers was by no means universal is apparent from Binning's description of a method of hauling guns muzzle foremost. He says:

> This will be thought a new Invention, but I used the same in my Lord *Middleton*'s Service from *Aberdeen* to *Fyvie*, where I caused them to make these Sled-feet, as you see fast to the carriage, in this manner; near to the Breech of the Peece there is a Bolt, whereon the end of the Sled-foot is; and under it, at the foot-end of the Carriage, a Square-hack to lay over the Sled-foot, and then a Rope through the Sled-foot; and a man or two thereby shall steer a Gun by a Height or Hole, in the way where she is drawn, so that many times it saves the Guns falling over. And when you are to meet your Enemy, or make use of your Guns, you may lift up your Sled-feet, and lay them all along the side of the Carriage in manner as you see, on a Hack where they do not trouble, and unhacking the Ropes from the Hacks before, you may use your Gun at your pleasure.

From an illustration it appears that the 'sled-foot' was a wooden bar which ran along each side of the trail and projected beyond it to raise the end of the trail from the ground. The 'square hack' must have been a metal bracket or hook. Through the end of each sled-foot was a rope, held by a man, by means of which the gun could be slewed round as required. This was obviously a much more cumbersome method of moving a gun than the limbers, or fore-carriages, which were now coming into general use.

Binning has also some advice on the loading and firing of mortars:

> Now he that would Load a Mortar-Peece, may elevate her Muzzle to what degree he will for his own con-

veniency; the Peece made clean, you put the Powder in the Chamber, and upon the Powder a Wad of Rope-yarn, Hay, or what you can provide; then you put a Turf of Earth cut on purpose, that is large, wider than the vacant Cylinder upon the Wad, which fills the Chamber, and then you put the Granado or other Fire-Work above that Turf, and putting Grass or Hay above your Granado, that it may lie as you would in the Mortar, and also to keep the Powder in the Mortar from the fire of the Feusee . . . . When you would discharge a Mortar Peece, first you must set fire to the Feusee of the Granado or Fire-Work, and you must see it burn well before you give fire at the Touch-hole.

It is interesting that Binning favoured the highly dangerous method referred to in the last chapter.

Sir Jonas Moore, in his *Modern Fortification: or Elements of Military Architecture* (1689), gives some interesting particulars of the ordnance of his day. The most important of these were as follows:

(a) There were three types of culverin: whole-culverin (also called dragon-drake), culverin, and demi-culverin. Of these, the first fired balls of 40–60 lb, the second of 20–35 lb, and the third of 14–18 lb. (These were considerably larger pieces than those of the same name used in the Civil War.)

(b) The thickness of the metal varied, and these variations were known respectively as small, common, and reinforced or fortified.

(c) The culverin varied in length, and was called accordingly extraordinary, ordinary, and bastard. The first, from touch-hole to muzzle, was 39–41 calibres long, the second 32 calibres, and the third 26–28 calibres.

(d) The field pieces also varied in calibre-length. They were the saker (or quarter-culverin) of 8–12 calibres, the falcon (or half-saker) of 5–7 calibres, the falconet of 2–4 calibres, and the smeriglio or

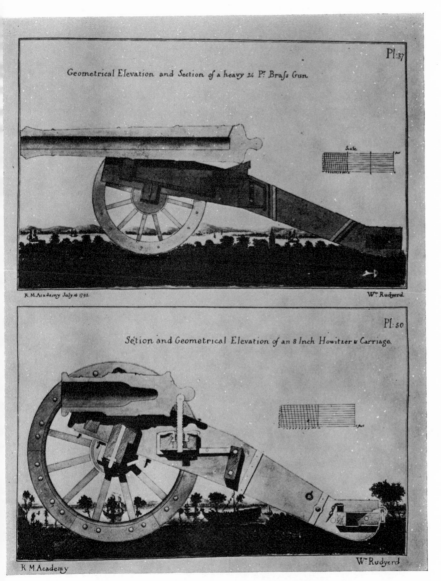

Geometrical Elevation and Section of a heavy 24 P.ʳ Brass Gun.

Pl. 37

Scale.

R. M. Academy July 16 1791.

Wᵐ Rudyerd.

Section and Geometrical Elevation of an 8 Inch Howitzer & Carriage.

Pl. 50

R. M. Academy

Wᵐ Rudyerd

1. A 24-pr brass gun (above) and an 8-in howitzer, ciraca 1790.

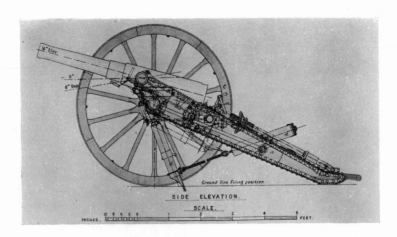

2. 9–pr BL field gun, circa 1911.

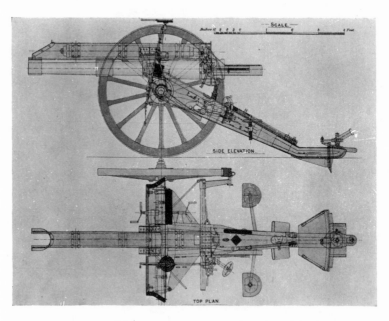

3. 15–pr BL gun converted to take a recoil buffer.

robinet of $\frac{1}{2}$–1 calibre. (These were also larger than
their predecessors.)

(e) 'Cannons of battery' were pieces ordinarily shorter
than culverins. They comprised whole-cannon of
70–100 lb calibre, cannon of 30–50 lb, and quarter-
cannon of 16–18 lb. (These were very similar in
size to pieces in this category used in the Civil War.)

An interesting development in the classifying of guns is
reflected in the first ships' 'draughts', of which there were
seven, dated 1684. The guns for the fourth-rates, for
instance, were given as culverins (12-prs) on the gun deck,
6-prs on the upper deck and sakers (5$\frac{1}{2}$-prs) on the quarter
deck. This showed that, while the old names were still
used, classification by weight of shot was coming in.[12]

At some time between 1665 and 1680 the proof of
ordnance was moved from Moorfields to the naval depot
at Woolwich. Two officers, a proof-master and 'His
Majesty's Founder of Brass and Iron Ordnance' (what a
magnificent title!), were appointed to supervise the construction of ordnance and to advise the various contractors,
who at this time manufactured all the ordnance for His
Majesty's naval and military forces.

Artillery played a large part in the campaigns in Ireland
of 1690–1. The best contemporary accounts of artillery
affairs are contained in *The Danish Force in Ireland 1690–1691*,
a magnificent collection of Danish documents very ably
edited by Messrs K. Danaher M.A. and J. G. Simms PH.D.
and published by the Irish Manuscripts Commission.
William III's army started the campaign with the English
artillery in a bad condition. Writing to King Christian V
from Belfast on 13 March, 1690, Lieutenant-General
Ferdinand Wilhelm, Duke of Württemberg-Neustadt and
Commander-in-Chief of the Danish contingent, said that
there were only three 'battering pieces' and one mortar.
He added, however, that artillery was expected from
Holland. In a further letter of 19 April, he explained that
the few battering pieces that had been used in the reduction
of Carrickfergus had subsequently been worn out in

training gunners. The Dutch artillery arrived on 16 June, 1690, and at the Battle of the Boyne a month later artillery was used extensively by both sides.

The most noteworthy exploit connected with artillery was at the siege of Limerick. In a diary of events, which he sent to King Christian, Württemberg complained that on 10 August the Jacobite defenders of the city were causing great damage to the Williamites with their guns. William III was, however, awaiting the arrival of his siege train to batter down the defences of Limerick. But on the evening of 11 August Württemberg heard that the Jacobite cavalry commander, Major-General Sarsfield, had crossed the Shannon with 500 horse and 60 dragoons to attack the siege train, which was marching from Dublin with an escort of only one squadron of Villiers' Regiment (The Bays) as escort. Major-General Lanier was therefore ordered out with 600 horse and 200 dragoons to forestall Sarsfield. He was too late. On 12 August news reached the Williamite camp that Sarsfield had surprised the siege train while the men were asleep. In the train there were 60 ammunition wagons loaded with 12,000 lb of powder and 3,000 cannon balls, eight 18-pr guns, a great quantity of match, grenades and carcasses, and a number of bread wagons with three days' supply for the army. The Jacobites blew up two guns and burnt all the wagons, gun-carriages, ammunition, supplies, and equipment. They then rode away with all the 500 artillery horses and three standards.

This Jacobite raid was a disaster of such magnitude that it led eventually to the raising of the siege of Limerick. The remainder of William's heavy artillery, amounting to 14 'battering pieces', was at Carrick, 40 miles away, having recently arrived at Waterford by sea and been ferried up the Suir. There was also 60,000 lb of powder, 6,000 tools, and 1,000 cannon balls for each gun. But all this material was completely immobile and constituted a large store dump. William asked all officers who had wagons and carts to send them to Carrick-on-Suir to try and bring up some of the guns and ammunition. But it was estimated that it would be eight days before any of it could arrive. (In fact, the difficulties must have been too great, because it does not appear that anything from the Carrick dump arrived before Limerick. It may be that shortage of horses was the main difficulty.)

On 16 August the remaining six pieces of the siege train were brought into the Williamite camp and mounted on carriages, but there was little ammunition for them. Rain came to the assistance of the defence and on 29 August it was decided to raise the siege because of the shortage of gun ammunition. (The last convoy had brought, instead of cannon balls, 'unfilled bombs', which were of no use because there was only one mortar.)

On 17 October Württemberg sent to Christian v his journal covering the events at Kinsale and the capture of the two forts which defended this important port. One fort was captured quickly, on 3 October. The defenders retired to the keep, 'where there were still twelve iron guns', but soon surrendered, and forty-six guns, including a bronze 48-pr, fell into the hands of the victors. The second fort was a much more difficult proposition and was stoutly defended. It was necessary to wait for the siege train; and this, owing to contrary winds, had had to be disembarked at Cork and brought overland to Kinsale, a distance of twenty miles, or what Württemberg calls 'ten long Irish miles'. Its movement was difficult, because there was a serious shortage of artillery horses. All officers, therefore, were asked to surrender wagons and horses; but only the Danes could help, because the English were travelling light and had neither horses nor wagons. The artillery train which was awaited consisted of eight 24-prs, eight 18-prs, six 12-prs, eight 9-prs, and thirteen mortars. On 11 October six of the 24-prs and two mortars arrived, and were in action the following day. The 24-prs were Dutch and the mortars English. Both made good practice, and Württemberg commented on the 'extraordinarily good Dutch gunners'. The remaining two 24-prs arrived on 12 October. On the 14th the mortars ran out of ammunition, and the English troops carried on the bombardment with 12-prs, firing only at the breast-works for they were far too light to make a breach. But that afternoon the Jacobites surrendered and, in recognition of their gallant defence, were allowed to march out with main and side arms to Limerick. Together with the fort, the Williamites acquired 94 guns, of which 34 were 'brass'.

The following year, in June, 1691, Athlone fell after a bitter struggle. Thirty guns were used in the attack, nearly all 24- and 18-prs. At the Battle of Aughrim in the following month the Williamite artillery comprised 12-, 9-, and 6-prs; these being apparently the usual field guns. An adequate siege train was available for the second attack on Limerick. Writing to Christian v on 23 August, 1691, Württemberg said that they had much better artillery than in the previous year with 1,000 rounds for each gun. In addition to the heavy guns there were 10 mortars and 4 howitzers.

The howitzer was a new weapon which had suddenly become important in a curious way. It was basically a mortar with an easily adjustable elevation; but it became much more valuable as an anti-personnel weapon owing to the discovery by Marshal Vauban at the end of the seventeenth century of the effective use of the ricochet. The howitzer, which was normally used like the mortar to fire fused shells at a high angle, could be set at a low elevation to fire against troops in the open. The shell would strike the ground short of its target and would then bounce and roll along the ground to explode in the opposing ranks.[13]

At this time there was no equivalent of the later battery organization of the artillery. The ordnance to accompany an army was formed into a field train. This might be divided into sub-units for administrative purposes but most of the guns were distributed amongst the infantry at two per battalion for fighting and manoeuvring. Sometimes the pieces stayed with these battalions on the march and in camp, but normally the train marched and camped as ac omplete unit. Officers and men were appointed by warrant to accompany the trains, and in 1693 they were organized into companies. In 1698 a regimental train consisting of four companies and a staff was established as a permanent part of the British Army.[14]

CHAPTER V

Frederick the Great

A T the beginning of the eighteenth century the most efficient artillery in Europe was that of the Austrian Army; and it was the heavy losses suffered by the Prussians from this artillery which caused that remarkable military genius, Frederick the Great, to revolutionize his own artillery arm.[1]

When King Frederick William I ascended the throne of Prussia in 1713 he started almost immediately to carry out a considerable reorganization of the army. Between 1726 and 1734 he organized the artillery into two battalions, one field and the other garrison, and amassed a considerable quantity of material to provide field and siege trains. His ordnance, cast in Berlin, consisted of 3-, 6-, 12-, and 24-prs, all of which were sufficiently mobile for use in the field. His artillery officers were excellent, but the army as a whole was more of a parade ground than a fighting force; spoilt by the King's insistence on drill and turnout at the expense of training in weapons and tactics.

Frederick the Great succeeded his father in 1740. The following year he was nearly defeated at the battle of Mollwitz. The battle was saved by the steady Prussian infantry and the accurate shooting by the heavy field guns distributed along their front. Frederick saw immediately that he must have more artillery and formed a second field battalion. At the time he thought that this was enough, but as he learned more of the value of artillery, so he increased the proportion of this arm.

Frederick wrote much on military affairs and the best of his writings have been very ably edited by Mr Jay Luvaas in his *Frederick the Great on the Art of War.*The great general is not always easy to follow, because he changed his mind on many matters in the light of experience.

In 1756, at the beginning of the Seven Years' War, Frederick increased his artillery strength from two to six field battalions, each of 900 men, and from one to two garrison battalions, which were distributed by companies in the various fortresses. Later he could have done with more, because, owing to the deterioration of his infantry after years of campaigning, he was forced to rely increasingly on artillery as the principal arm. Nevertheless, there was a limit to the size of the artillery because of the enormous number of horses which it required in first line and reserve.

Unusually for this period, Frederick insisted that weapon training and exercises were to be given considerable attention during times of peace. All his artillery battalions were housed in large barracks which had been specially built for them in Berlin; and in the neighbourhood firing was carried out in the spring, followed by separate exercises for field and siege artillery. Field artillery was exercised in both attack and defence, and, in the former, gunners were expected to manhandle the guns forward themselves, keeping up with the infantry advance and giving rapid and accurate covering fire.

Ordnance was expensive and the Prussian artillery became a heavy burden on the State. But Frederick made sure that it was always equipped with the latest and most efficient weapons. For instance, he acquired seventy 10-pr howitzers which would fire grenades to an altitude of 4,000 feet, and which he considered would be invaluable in attacking positions on mountain heights which could not be reached by guns.

Frederick thought that artillery was most troublesome on the approach march. Near to the enemy, guns moved with the brigades to which they had been allotted; but under other conditions the slow-moving artillery train had to have infantry battalions detailed for its protection. For its mobility the artillery depended, of course, on its horses and special measures were taken to look after their welfare. The men normally engaged as drivers were those who were useless for any other military employment, and of a type which could not be trusted to do any job properly. To ensure that the horses received proper care, therefore, Frederick posted to the train for this purpose officers too old for service in the field and old N.C.O.s, promoted to subaltern, to assist them.

The roads of the time posed frequent difficulties to the movement of artillery. In spring and autumn, rutted and muddy roads, combined with the shorter days, restricted the length of a day's march to fifteen miles. Good roads were so important that Frederick laid down that, even if the best roads entailed a longer march, they must be taken and that artillery should follow the column which had been allotted the route most suitable for wheeled vehicles. If detachments had to be sent off any distance, he directed that they should not be accompanied by guns heavier than 6-prs, because of the difficulty of hauling heavier pieces with their associated equipment and ammunition.

Frederick's instructions as to the use of infantry guns were precise. Whilst the army was deploying, these guns were moved forward in front of the line to give covering fire. They then dropped back into line, and during the subsequent advance sufficient interval was left between adjacent battalions for two infantry pieces to move forward between them. The second infantry line was formed 300 yards in rear of the first and had only half the number of pieces in line – one between each adjacent battalion.

The allotment of ordnance is interesting. Each battalion in the first line was given two 6-pr howitzers and one 7-pr gun; and a heavy battery of ten 12-pr guns was allotted to each five-battalion brigade. The second line battalions had only two long 3-pr guns each, but the allotment of 12-prs to brigades was the same as for the first line. The 10-pr howitzers, of which there were forty, were held in reserve.

Frederick's great innovation was the introduction of horse or, as he called it, light artillery. This consisted of 6-pr guns and howitzers mounted on well-built carriages and drawn by specially selected teams of six strong horses, driven postilion fashion by three drivers. The gunners, numbering seven or eight, were all mounted on Polish horses and each small battery of six pieces was commanded by a captain assisted by two subalterns. One of the gunners acted as horseholder. In the limbers were carried 80 rounds of shot and 20 of canister per gun. The first battery was formed in 1759 to support cavalry in detached operations. Eventually there were four batteries, totalling twenty 6-pr guns and four howitzers.[2]

Frederick emphasized the importance of artillery in the attack, and said that the bulk of it should be on the attacking wing. If there were suitable hills the heaviest batteries should be placed on them to bring the enemy under a cross-fire to breach his defences. If there were no such hills, howitzers would have to be used instead of guns. At the start of the attack the light field pieces attached to infantry battalions should open continuous fire when within 600–700 paces from the enemy. At 500 paces horse teams should be sent to the rear and the guns moved forward by the men, in such a way that a continuous fire could be kept up. The primary target in the attack was the enemy guns to stop them firing on the advancing infantry. Each of these battalion pieces was served by a team of six gunners and three carpenters. If the enemy began to retreat the heavy guns should be brought forward to harass them.

Frederick pointed out that close-range fire was far more effective than shots at long range. Not only would a cannon ball penetrate all the enemy lines, but the noise of a shot passing overhead was terrifying, and thus very effective in reducing enemy morale. He therefore forbade the premature opening of fire at long range because, he maintained, it only wasted ammunition and had no effect.

He took some time to make up his mind on canister, but eventually directed that it should not be used at ranges greater than 100 paces, because at longer ranges the pellets became too scattered to be effective, some of them striking the ground in front of the enemy and others flying over their heads.

Against a cavalry attack, the guns should open fire at a maximum range of 800–900 paces and as rapidly as possible. The switch to canister should be made only in sufficient time to deliver the first volleys at 60–50 paces. To keep up a continuous fire, half a battery was to fire at a time and there was to be no fire by individual guns.

In defence the main target was the enemy infantry, and all guns were to concentrate on smashing the advancing lines. Ricochet fire should be used by the 6-prs until the enemy was within 100 paces when guns should start shooting canister. The first-line batteries should be laid out to fire obliquely and cover the whole front with cross-fire. Small howitzers firing shell were particularly effective. The guns of the second line did not come into action until those of the first line had been forced to fall back.

In withdrawal the large guns were to be sent away first, and if this entailed the descent of a steep hill the field guns should also be sent back, as otherwise, if they overturned on the descent, there might be a risk of their falling into the hands of the enemy.

Guns were not to be sited on too high hills, because the steep angle would prevent their shots from having such a penetrating effect. Furthermore, if a shot struck soft ground in front of the enemy it would remain stuck in it; if it struck hard ground it would bounce over their heads. When the terrain allowed, ricochet fire should be used against enemy infantry, as it was by far the most effective. When one's own infantry was advancing to the attack, one should never if possible fire over their heads, because they would duck at each shot and this would slow down the advance.

Frederick urged the use of mortars in the field. In an attack against a defensive position, ten mortars should be formed into two batteries for counter-battery bombardment during the infantry advance.

The influence of Frederick's innovations was first felt abroad in the French artillery. French ordnance had been standardized in the second half of the seventeenth century into five different calibres: 4-, 8-, 12-, 24-, and 32-prs.

These were fixed in 1668 and confirmed and modified in 1732. After the Seven Years' War, and following a series of comparative trials at Strasbourg in 1764, Gribeauval, the first inspector of artillery, began a reorganization of the French artillery. To obtain a badly-needed mobility he selected 4-, 8-, and 12-pr guns, all plain unchambered pieces of 18-calibre length, and 6-in. howitzers. All these fired well-fitting projectiles with powder charges about one-third the weight of the shot. Limbers were large six-wheel vehicles; six-horse teams hauled the 12-prs and four-horse teams the rest. All other artillery was classified as siege or coast defence. By the end of the eighteenth century all the other great powers had adopted Gribeauval's classification.[3]

The British Army's debt to Frederick the Great was shown by the formation in January, 1793 of two bodies of Horse Artillery, each called a troop and having four (later six) guns.[4] Horse artillery was adopted by the French Army at about the same time (after the outbreak of the Revolution).[5]

CHAPTER VI

CHAPTER VI

British Ordnance of the Eighteenth Century

A remarkable man, Benjamin Robins, was born at Bath in 1707. Although a Quaker, he became the foremost artillery expert in Europe. In 1742 he published a book on gunnery in which he explained the value of rifling and the advantage of using elongated projectiles.[1] In 1743 his paper on *New Principles of Gunnery* was read before the Royal Society. In this he dealt with both internal and external ballistics, showing that air resistance was vastly greater than had been generally thought. A little later he addressed the Royal Society on the importance of rifled guns and urged the adoption of rifled ordnance and fire-arms. He was unable, however, to get his advocacy of rifling taken up.[2]

In 1747 Robins produced a pamphlet on the more efficient disposition of metal in ordnance. He pointed out the advantages of large shot over small shot, in giving increased range and, at sea, greater penetration with a bigger hole. A ship, he deduced, should be armed with the largest guns she could safely bear. It followed that pieces should be made as light as possible by cutting out all metal that was not necessary to strengthen a piece. He recommended reducing the powder charge to one-third of the weight of the projectile, irrespective of the calibre of the piece. This smaller charge, he said, would give an almost similar range to the heavier charges currently in use and would be much less harmful to the gun.[3]

Somewhat similar views were expressed by another great artillery theoretician, John Muller. Muller, a German, was born in 1699, but came to England at an early age and in 1736 wrote his first book while apparently working at the

Tower of London. In 1741 he was appointed headmaster of the Royal Military Academy, Woolwich, and when this establishment was reorganized, he was appointed professor of fortifications and artillery, retaining this post until his retirement in 1766. In 1757 he wrote the first edition of his *Treatise of Artillery*, of which a second edition appeared in 1768, and a final and enlarged edition in 1780, four years before his death. He wrote many other works on mathematics, fortifications, field engineering, gunnery, and so forth.

Muller also not only believed that charges were too great and pieces too heavy, but carried out experiments to support his views. He declared that adherence to tradition was so tenacious that only the express order of the Duke of Cumberland could overcome the objections to the introduction of light field pieces into the British Army. As regards the proper length and charge for a piece, he drew the following conclusion from his experiments: 'If the length of a piece of any calibre be 21 diameters of its shot, and loaded with powder equal to half the weight of the shot, it will carry farther than any other of the same calibre, either longer or shorter, loaded with any charge whatever.'

He appreciated, however, that conditions of service prevented these ideal dimensions being realized in practice. Ship guns had to be short 'so as to be easily housed and loaded' and field pieces had to be light and short for mobility. For these reasons he thought that field pieces should be 14 diameters in length and ship guns 15 diameters; and he says that they had found one-fourth of the weight of the shot sufficient charge for field pieces.

Muller carried out experiments on mortars in conjunction with Colonel (later General) Desaguliers, the great French artillery soldier. He also held strong views on the superiority of iron guns over brass, and he adds: 'Although the artillery officers agree, that iron battering pieces are preferable to brass, yet to make field pieces of iron they by no means approve of, because they say it would be too dangerous to stand by them in time of action; but what should prevent a proper trial being made?' Muller's book gives a fascinating account of the changes he considers necessary, set against a description of contemporary practice.

Something of a revolution in the art of casting ordnance occurred in 1739 with the invention at Geneva by the Swiss gun-founder Maritz of the boring machine. By using a lathe driven by horses Maritz was able to bore guns from the solid casting. He was invited to demonstrate his invention in France and set up a plant at Lyons, followed by another at Strasbourg. Not only was it now much easier to produce a true bore, as compared with the previous method of casting the piece hollow on a core, but iron guns could be made much stronger. Since their cost was only about one-eighth that of bronze, they increased considerably in favour.[4]

Description of the Cast Piece

Smooth-bore muzzle loaders settled down in the eighteenth century to a standard method of construction and nomenclature. The muzzle was shaped rather like the conventional chimney of a railway engine, being flared out to a lip in what was known as the 'swell of the muzzle'. In front of this the 'muzzle mouldings' tapered inwards to the bore. A short distance behind the swell of the muzzle

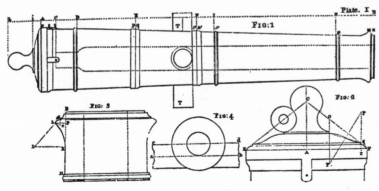

The parts of a gun, circa 1780.

there was a raised band round the barrel, flanked by two lesser bands, and known respectively as the 'muzzle astragal and fillets'. Rather less than half-way back along the length of the piece were the 'chase astragal and fillets', and a little way behind them the 'second reinforce ring and ogee' – a combined moulded band, tapering down to the barrel towards the muzzle. From this band to the muzzle was known as the 'chace'. As the name of the band signified, it was the end of the 'second reinforce'. The major part of the barrel was divided, in fact, into the first reinforce, the second reinforce, and the chace. 'Reinforce' signified thicker metal. The second reinforce had a thicker belt of metal round the bore than the chace, and the metal of the first reinforce was thicker than that of the second reinforce. The idea was that the bore should be surrounded by progressively less metal as the force of the explosion lessened towards the muzzle. The next band back, then, was the 'first reinforce ring and ogee'. Farther back, and just in front of the vent, was the 'vent field astragal and fillets'. At the rear of the barrel was the 'base ring and ogee', and between this and the vent was what was known as the 'breech'. At the rear, behind the breech, was the cascabel, consisting of the breech mouldings tapering down to the 'button', or knob.

The trunnions were placed just in front of the second reinforce ring, so that there should be rather more weight behind than before them, with the object of preventing the gun from kicking up behind when it was fired; which it would otherwise do, because of the traditional position of the trunnions below the centre line of the piece. The handles on the barrel for lifting the piece were called 'dolphins' (because they were generally shaped as such) and were so placed that when the gun was suspended they were at the point of balance.

The relative lengths of the different parts of the gun were established by convention. The length of the gun (less the cascabel) was considered as divided into seven equal parts. Of these, two parts were allotted to the first reinforce, one part plus the diameter of the bore to the second reinforce, and four parts minus the diameter of the bore to the chace. The length of the cascabel was 2¼ calibres, and the length of the breech equalled the thickness of the metal at the vent. The trunnions were always a calibre in length and in diameter, and so placed that a line drawn through their centres would touch the lower point of the bore.[5] The function of the button at the back of the cascabel was to hold one end of a sling for lifting the piece, the other end being secured to a handspike inserted in the bore.

Mortars and howitzers were constructed in a rather different manner to guns. A mortar was a short piece with a large bore and a chamber, and was primarily intended to fire shells from a high angle. The principal parts of the mortar were the reinforce (which was a wide belt about the middle of the piece with dolphins on it), the chase (which was the short length of barrel between the reinforce and the barrel), and the breech (which was the rounded end behind the reinforce). The trunnions were at the breech end of the piece.[6] Muller writes:

> Of all the parts of artillery, the construction of mortars is the most variable and uncertain; almost every artillerist has some favourite notion or other concerning their figure. . . . The parts of mortars are formed in imitation of those of guns; for which reason they make them with a reinforce. This only overloads the mortar with a heap of useless metal and that in a place where the least strength is required.

The howitzer, or 'howitz' as Muller calls it, was somewhat akin to the mortar in construction, but it had its trunnions in the middle of the piece, and, unlike the gun, they were level with the axis of the bore. Like the mortar, the howitzer was intended to fire shells. But whereas the mortar had a fixed quadrant elevation of 45° (the range being adjusted by altering the weight of the charge), the howitzer could be fired at variable elevations and horizontally against troops in the open. They were far more mobile than mortars but were approximately twice as heavy for the same bore.[7]

The Carronade

Robins's paper of 1747 had impressed a very able gunner, Lieutenant-General Robert Melville, and in 1774 he advocated a short piece of 8-in. calibre firing a 68-lb ball, but with a powder charge of on' ˜ 5½ lb. He persuaded the Carron Company, a Scottish ˙ ⌐n-founding and shipping firm, to cast such a piece, which he named 'the Smasher'. The Company were impressed and at its monthly meeting of December, 1778, the Manager told the Board that he had constructed a very light species of gun, resembling a coehorn, to provide armament for the Company's sailing packets. Various people, he said, had been impressed by it and, if the Board agreed, he could get a quantity of orders. The Board did agree and decided that the new piece should be called a 'carronade'. The Carron Company later on presented Melville with a model of this most successful venture inscribed 'Gift of the Carron Company to Lieutenant-General Melville, inventor of the Smashers and lesser carronades for solid, ship, shell, and carcass shot, etc.: First used against French ships in 1779.'

With the conventional long gun it had become difficult to sink a ship by gunfire, because the holes made by the projectiles were too small. The big carronade made a large irregular hole, and the windage had been cut down considerably by Melville so that the small charge of powder achieved far more than would have been possible with the normal pieces of the time. Even so, the carronade's range was short and its powers of penetration were poor. But for close-range fighting, particularly at sea, it was magnificent, and carronades became so popular in the Royal Navy that by January, 1791 they had been mounted in 429 ships.

The prototype had trunnions, but subsequent pieces were cast with lugs, by which they were attached to wooden slides, recoiling in slotted carriages.

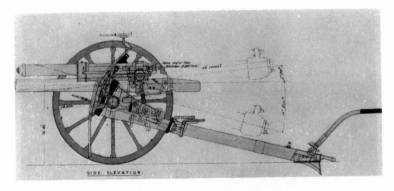

4. 13-pr QF field gun.

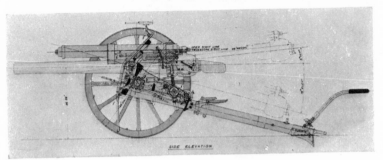

5. 18-pr QF field gun.

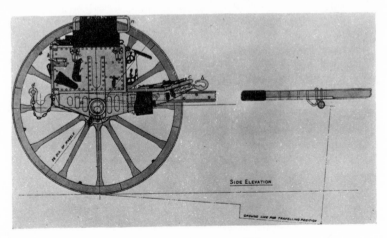

6. Limber for the 18-pr QF gun.

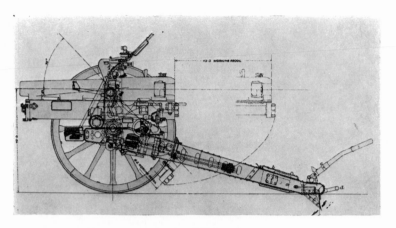

7. 4.5–in. QF howitzer.

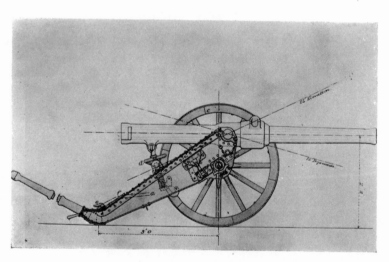

8. 10–pr BL mountain gun.

The Eighteenth Century Artillery Train

Muller provides interesting details of the field artillery trains in the middle of the eighteenth century. For the campaign of 1747 there were the following:

Item		Number	Horses, each	Horses, total
Kettle Drum		1	4	4
Tumbrils		2	2	4
12-pr Guns		6	15	92
9-pr Guns		6	11	66
6-pr Guns		14	7	98
3-pr Guns		26	4	104
Howitzers (8-in.)		2	5	10
Ammunition Carts		20	3	60
Forge Carts		2	2	4
Pontoons		30	7	210
Spare Carriages,	Pontoons	3	7	21
"	12-pr Gun	1	7	7
"	9-pr Gun	1	5	5
"	6-pr Gun	2	5	10
"	3-pr Gun	4	3	12
Spare Limber,	12-pr Gun	3	2	6
"	9-pr Gun	3	2	6
"	6-pr Gun	2	1	2
"	3-pr Gun	3	1	3
Spare Horses				20
			Total	744

Baggage wagons for officers and wagons for ammunition and stores added another 795 horses. Muller mentions that the 'flag gun, which is a 12-pounder, had 17 horses to draw it, although all the rest had but 15, which makes up the number of 92 horses for the six 12-prs: with regard to the waggons, some were drawn by three horses and others by four.'

Muller gives the 'Order of General Belford's march of

the Artillery', which must refer to the first Continental campaign of the Seven Years' War. William Belford (1709–80) was a Lieutenant-Colonel in the Royal Artillery in 1749, a Major-General in 1758, Commander Woolwich District in 1758, and General in 1777. Belford's order of march was as follows: a guard of the army; the company of miners with their tumbril of tools drawn by two horses; the regiments of artillery front guard; the kettle drums drawn by two horses and two trumpeters on horseback; the flag gun drawn by seventeen horses and five other 12-prs each drawn by fifteen horses; eleven wagons with stores for the 12-prs and one spare wagon drawn by three horses each; six 9-prs each drawn by eleven horses; nine wagons with stores for the 9-prs and one spare wagon, each drawn by three horses; five long 6-prs each drawn by seven horses; seven wagons with spares for these guns and one spare wagon with three horses for each; five more long 6-prs with the same number of horses as before; seven store wagons, including one spare, each with three horses; four long 6-prs with seven horses apiece; six store wagons, including one spare with three horses for each; two howitzers, each drawn by five horses; four wagons, each drawn by three horses, with howitzer stores; six short 6-prs, each drawn by two horses (note the tremendous difference as compared with the long 6-prs); three 6-pr store wagons each with three horses; six royals and their stores with three horses apiece; one 12-pr carriage drawn by nine horses; one 9-pr carriage with five horses; a long 6-pr carriage with five horses; two short 6-pr carriages each with two horses; one short and one long limber hauled by one horse each, and two forges with two horses each; twenty ammunition carts each drawn by three horses; nineteen wagons with musket cartridges and one spare wagon each drawn by three horses; thirty powder wagons and one spare with three horses each; twenty-five wagons with entrenching tools and one spare wagon, each pulled by three horses; twenty-five wagons and one spare wagon for small stores each with three horses; six wagons for artificers with four spare wagons – the large

Transom Bolts { Fore.......a
             { Center....b
             { Trail.....c
Trunion Plate..........e
Cap Square.............f
Joint Bolt.............g
Eye Bolt...............h
Bead Plate.............k
Transom Plate with Hooks...l
Side Steps.............m
Locking plate.........p
Lashing Rings.........q
Axletree Band.........r
Garnish { Bolts.......s
        { Nails.......u
Washer................n
Linchpin..............

draught Rings.........w
Single forelock key...x

6-pr gun carriage.

number of spares is curious – hauled by three horses each; thirty-two baggage wagons of which nine were hauled by four horses and twenty-three by three – note, no spare wagons – thirty pontoons and three spare carriages, each pulled by seven horses; the artillery rear guard; the rear guard of the army. Marching with the guns were parties of gunners and matrosses, and pioneers were interspersed at intervals to repair the shocking roads where this long line of heavy vehicles had damaged them.

Employed in this campaign were 1,415 horses, 32 guns, 2 howitzers, 6 small mortars (the royals), 244 wagons and carts, and 30 pontoons. It was considered that 20 pontoons would suffice for any part of Flanders, because no river needed more to bridge it; there was therefore a useful reserve.

The above details present a useful picture of a British field artillery train on the march in the middle of the eighteenth century. The pieces of ordnance listed were all 'brass'. The 9-pr, rather oddly, went out of use after 1750. The 'long 6-pr' weighed 12 cwt and was 7 ft long; whereas the 'short 6-pr', known as 'General Belford's', weighed only $5\frac{1}{2}$ cwt and was 5 ft long. The 8-in. howitzer weighed $12\frac{3}{4}$ cwt, and since it required fewer horses to haul it than the long 6-pr, it must have had a far lighter carriage. The 'royals' were $5\frac{1}{2}$-in. mortars.

On the roads of the time, as already remarked, the haulage of ordnance was a problem. Muller writes:

All the carriages made use of in the artillery have shafts; and to prevent the great length of those that require a great number of horses, the rule is to draw by pairs a-breast, which is an absurdity no where else to be met with; for when the road is frequented by carriages drawn by two horses a-breast there is always a ridge in the middle, which the shaft-horse, endeavouring to avoid, treads on one side, whereby the wheels catch against the ruts, and stop the carriage; and when the fore horses bring them back, he treads on the other side, where the same happens again; so that the shaft-horse, instead of being useful any other ways than to support the shafts, becomes a hindrance to the rest.

It was not only for the horses that the ruts presented a difficulty.

> The span, or interval between the wheels, varies in different countries; even every county in England observes a different width, which is very inconvenient for those who travel in carriages. The artillery carriages are made like those in Flanders, which is four feet eight inches.

If all counties had used vehicles with the same span, long distance movement over the rutted roads of the period would have been much easier. The span used in Flanders may have been adopted with an eye to easy movement in Continental campaigns. It is conceivable that it then became something of a standard in Great Britain; but it is interesting that the gauge which became the standard for railways, first in Great Britain and then in the United States and most of Europe, should owe its origin to the Royal Artillery.[8]

Carriages

Gun carriages in the eighteenth century were of the so-called 'double-bracket' type; that is to say, the trail was formed of two wooden planks called 'cheeks' or 'brackets' which were placed on edge parallel to each other and connected by four transoms (cross-pieces), called respectively the trail transom, the centre transom, the bed transom, and the breast transom. The axletree was just behind the foremost of these, the breast transom. The bed and centre transoms were close together and over them was a board called the bed. On the top of the trail, and over the forepart of the axletree, were the trunnion holes (formed by semi-circular holes cut in the cheeks), and 'capsquares' (curved iron straps) fitted over the trunnions to hold them in the trunnion holes. On top of the bed were placed the quoins,

69

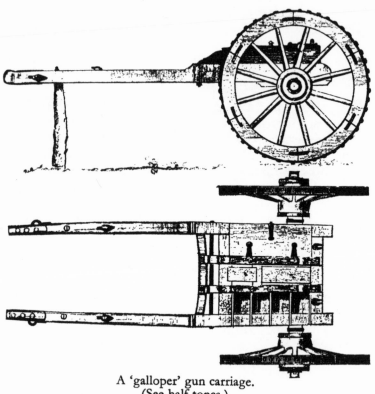

A 'galloper' gun carriage.
(See half tones.)

or wedges, on which the base ring of the piece rested and which, by being moved forward or backward, raised or lowered the breech. In the trail transom was the pintle hole, through which went the pintle, or vertical spike, of the limber. The pintle was usually on top of the limber axletree.

During this period the limber still consisted of only an axletree, wheels, and shafts. The shaft horse, according to Muller, had a wretched time.

The wheels of the limber being but four foot high and the extremities of the shafts five, the draught of the shaft horse becomes so oblique, that the greatest part

70

of his force is lost in supporting the fore ends of the shafts, which the other horses draw down again, so as to bring the whole draught in a right line from the axletree to the breast of the fore horse; whereby the shaft horse is so shook (the difference between the height of the fore end of the shafts and the centre of the axle-tree being at least two feet) that he is spoilt in a short time.

What is remarkable, however, is that Muller, not being able to see a solution to this trouble, suggested that the only remedy would be to fix a pole or shafts to the head of the gun carriage and draw it with trail sliding along the ground – a truly retrograde return to the laborious method of Captain Binning.

There was indeed one carriage which dispensed with a limber. This was the 'galloper' carriage which was used for very light pieces that could be hauled by one horse and fast enough to keep up with cavalry. Instead of a trail, there was a pair of shafts; and these, in fact, served as a trail when the gun was in action.

Numerous other vehicles formed part of the artillery train. A tumbril was a two-wheel cart to carry pioneers' and miners' tools. A powder cart also had two wheels, and contained shot lockers and space for four barrels of powder. An ammunition wagon was considerably larger, with four wheels, and Muller says, 'This waggon serves likewise to carry bread, it being lined round the inside with basket work.' The block carriage consisted of two long beams lengthwise on four wheels and was used to carry guns which were too heavy to be moved on their own carriages. A sling wagon was a four-wheel vehicle, with rack and handle mounted in the centre, for moving mortars and heavy guns over short distances from one position to another. The forge cart originally had two wheels, but it was so difficult to keep it steady that they were replaced by four. It was equipped with bellows, tool space, iron plate for the fire place, wooden trough for water, an iron plate for cinders, and another iron plate to stop the flame from setting fire to the cart. There was also a gin, a sort of portable crane for lifting ordnance on and off their carriages or mortar beds.

Ammunition

Cartridges came into increasing use during the eighteenth century, the requisite powder charge being wrapped in paper. Loose powder was retained for mortars, however, because the weight of the charge was adjusted to the required range.

The principal projectile was the round solid shot, but as its effect depended largely on a flat trajectory, it was normally fired only from guns, and constituted more than 70 per cent of their ammunition. A round shot was fastened to a sabot, which was a circular wooden plate, normally made of elm, and intended to seal the bore behind the shot to prevent the escape of gas. Case, or canister, was mainly used in defence and at up to 300 yards (instead of the 100 yards which Frederick the Great had finally decided was effective limit). Common (i.e., explosive) shell was fired from howitzers or mortars. The windage in a piece was sufficient to allow the flash from the charge to ignite the fuse. There was also an incendiary projectile known as a 'carcass'. A container made of layers of paper was filled with composition which was heated, poured into it in liquid form, and then allowed to harden. The container was pierced with a few holes so that the flash of the charge could ignite it. Carcasses were only fired from howitzers and mortars.[9]

The early eighteenth-century fuse was made of quick match enclosed in an iron tube, which was inserted in a hole in the shell casing. By the middle of the century the iron tube had been replaced by one made of beech wood. This wooden tube had rings cut round the outside, spaced half-a-second's burning time apart, so that the fuse could be cut to the appropriate length before placing in the shell.

CHAPTER VII

The Smooth-Bore Artillery of the Nineteenth Century

Napoleon

NAPOLEON was one of the most eminent in the line of great artillery officers, and his views on what the artillery of his time should consist of are worthy of attention. Whilst a prisoner at St Helena he dictated some notes on artillery to Baron Gourgaud, and some explanation of these notes was later added by Capitaine de Riviers de Mauny.

When the Revolutionary War broke out in 1792 the French field artillery was almost unique in being equipped with 4-pr and 8-pr guns. All the other European armies, except the Spanish, were armed with 6-prs, 12-prs, and 24-prs. One disadvantage of this from the French point of view was that they could not use captured ammunition; though of course the same applied to their enemies. When Napoleon became First Consul he decided that French ordnance should in future conform to these standard European calibres. His field artillery was to consist of 6-pr and 12-pr guns and 24-pr howitzers. For siege artillery there were to be 24-pr guns of two types, one short and one long. It was some time before the Army was equipped with these pieces, and indeed it was only during the campaigns in Russia of 1812 and Germany of 1813 that the Grand Army's artillery consisted entirely of the new equipments. But the heavy losses in these disastrous campaigns inevitably led to the reappearance of the old 4-prs and 8-prs, which were taken out of fixed defences to rearm the depleted field batteries. When Napoleon retired from the scene these old pieces remained in use under the restored Bourbon dynasty, primarily for economy.

In discussing his changes, Napoleon said that it had been right to replace the 4-pr and 8-pr by the 6-pr, because no officer could well decide which of the two former should be used, and even then – a touch of a somewhat Irish humour here! – he could only use what he had got. In any case, the frequent result was that 4-prs had to be used when 8-prs were needed or, alternatively, 8-prs were used when 4-prs would have sufficed. He deduced that a single calibre was sufficient for field use, provided that there was a heavier gun, a 12-pr, in reserve. As an exception, he considered that a 3-pr gun was useful as a mountain equipment. Guns of higher calibre than the above he considered useless in the field. As regards howitzers, he said that the 6-in. howitzer was too wasteful and he had therefore selected the 5½-in. (24-pr), which had the additional advantage that the wagon could hold 75 rounds for it as compared with only 50 rounds for the 6-in. He believed, however, that there should really be two types of howitzer, one to work with the 6-pr gun and the other with the 12-pr gun. The 10-in. mortar he abolished, but he kept the 6-in., 8-in., and 12-in. The 8-in. he regarded as excellent and the true siege mortar. The 6-in. mortar was, in his opinion, useful for the attack and defence of particular localities. It weighed less than 100 lb and was very effective against trenches and covered ways.

A horse artillery battery, Napoleon said, should have six equipments and a field battery eight. He would have preferred four for both, but this would have entailed almost doubling the number of artificers, spare stores, forges, etc. In each horse and field battery, he said, there should be two howitzers. This large proportion of howitzers was necessary for such tasks as dislodging an enemy from a village or shattering a redoubt. A field battery, then, would have six 6-pr guns and two 5½-in. howitzers. In vehicles it would have eight gun carriages in use, two spare gun carriages, a forge, a *prolonge* (divisional ammunition wagon), six ammunition wagons each hauled by from six to twelve mules, and four wagons for shells, each pulled by four mules.

British Artillery

The British smooth-bore period in the nineteenth century begins at the Napoleonic Wars and ends at the Crimean War and the Indian Mutiny.

The formation of horse artillery in the British Army is narrated in Chapter V, and these two troops of January, 1793 were followed by two more in the following November equipped with 6-pr guns. In the same year the Driver Corps was raised so that this mobile branch should be free from the iniquitous system of being dependent on hired horses driven by civilians.[1]

The organization of the Horse Artillery into tactical units was much more of a revolution than the introduction of the Horse Artillery itself. Hitherto the artillery for field service had been formed into a field train. The train was nominally divided into 'brigades of guns' of about twelve pieces each, but these pieces were distributed for action amongst the infantry, two being allocated to each battalion. The train was drawn by horses, with civilian drivers, which had been either hired or purchased for the campaign; and temporarily employed 'conductors' were responsible for their management. From 1794, sections of the Driver Corps were allotted to the Field Artillery, as required,[2] and the hired horses with their drivers and conductors began to disappear.

The field train included artificers and other necessary personnel on its establishment, but the artillerymen themselves were concerned solely with the service of their pieces. Thus, one artillery company of 100 gunners sufficed for a brigade of 12 pieces (that is, 8 gunners per piece). This organization was retained throughout the first period of the Napoleonic war and until its recommencement in 1802. It did not apply to the Royal Horse Artillery, the organization of which is described above.[3]

In 1802 the field artillery was reorganized into brigades of six pieces each. Throughout the Peninsular War and the subsequent Waterloo campaign the field artillery unit of six pieces was designated a brigade. The horse artillery tactical unit was still called a troop, but the whole of the R.H.A. was referred to as the Royal Horse Brigade.[4] In 1806 the Driver Corps was renamed Corps of Royal Artillery Drivers; in 1821 it was abolished and artillerymen were enlisted to serve as either gunners or drivers.[5] In 1919 the term 'field battery' was introduced by the Committee of Revision in replacement of 'brigades' for units which had been so described. But the term 'brigade' remained in common use for many years afterwards.[6]

As field batteries were not required to move faster than a walk, the gunners marched on foot, but they could be mounted on the carriages if necessary. In troops of horse artillery, gunners were, of course, all mounted or carried on carriages and wagons.[7]

In the Royal Artillery of 1854 there were seven troops R.H.A. and twelve battalions R.A.; and in each battalion there were eight companies. There was also an Invalid Battalion of old soldiers, usually employed in forts and other fixed establishments. A varying number of the companies were from time to time designated 'field batteries'. Such batteries were then either attached to instruction batteries at Woolwich or were moved to stations (generally somewhere in the United Kingdom) with equipments of various kinds. The instruction batteries held equipments and horses for several field batteries, and these were taken over by companies in turn for their period of training. Operational field batteries were called 'batteries of service'. The rest of the Royal Regiment of Artillery consisted of garrison companies stationed at home and overseas.[8] In April, 1859 it was directed that all troops and companies should be redesignated batteries.

Manufacture of Ordnance

By 1850, the last period of the undisputed sway of the

smooth-bore gun, cast iron ordnance were being made at the establishments of private founders who were furnished by the Ordnance Department with the necessary drawings. The principal gun-founders were the Low Moor Company, with works near Bradford, and Messrs Walker of Gospel Oak in Staffordshire. The Carron Company, which had been producing ordnance over a long period, had turned to other activities.

The first step in manufacture was to make a model of the piece, either in well-seasoned hard wood or cast iron. A cast iron jacket was then made for making the mould. It usually consisted of twelve parts, each having flanges at the edges. The flanges were perforated so that bolts could be inserted to connect the various parts. The whole jacket was lined with sand moistened with a solution of clay to make it adhesive; and it was then built round the model, which had itself been covered with charcoal dust, also moistened with the clay solution, and known as the 'blacking'. The object of the blacking was to stop the mould from sticking to the model. Sand, moistened in the same way, was then rammed between jacket and model to take the shape of the latter and thus form the mould. The jacket was then unbolted, removed from the model, and placed in a drying store for about twelve hours. When dry, the parts of the jacket were taken to a pit, joined together, and secured in an upright position. The melted metal was then poured into the top; and the resultant casting was left undisturbed in the pit for twenty-four hours. It was then taken out, prepared for the lathe, and placed on the boring bench. When the bore was completed it was removed to the turning bench. Finally it was vented.

'Brass' (i.e., bronze) smooth-bore guns were made in an entirely different way. A model of the gun, somewhat enlarged, was formed on a wooden mandril. To prevent adhesion, the mandril was covered with a substance consisting of straw plait on which had been pasted a mixture of moistened sand and horse-dung. The mould was then worked on to the model. It was built up by successive layers of loam mixed with cow hair and long fibres of hemp.

Each layer was partially dried over a charcoal fire before the next was applied. Strips of iron were then laid longitudinally over the mould and bound round with hoops. As the mould contracted during the drying process, wedges were driven under the hoops to keep everything tight. Finally the whole was covered with loam mixed with water and then dried. The smaller end of the mandril was now given a few sharp blows to remove it from the mould. The metal which was subsequently run into the mould consisted of 92 parts copper and 8 parts tin. The boring out, etc., followed.[9]

Types of British Ordnance

As in the French Artillery, the British field equipment most constantly in use was the light 6-pr, the brass piece being 5 ft long. After the battle of Waterloo it became the Horse Artillery equipment. The 'brass' 9-pr gun was introduced in 1808 and became the very effective weapon of the Field Batteries. The 'brass' 12-pr gun was not very popular for European warfare but was popular in India, where a bigger weapon than the 9-pr was needed to demolish the gates of the many minor forts. The light 5½-in. howitzer was a very short piece and consequently erratic; but it had a high trajectory and fired a powerful common shell for its size. The heavy 5½-in. howitzer was a very good and accurate piece. It was the practice, again like the French, to include one or two of these howitzers (either the light or the heavy version) in every troop of Horse Artillery and in every Field Battery. In 1822 they were succeeded by the longer 12-pr and 24-pr howitzers.[10]

In 1839 there were the following three types of Field Battery:

		Horses
(a)	*12-pr (Medium) Field Battery*	
	5 12-pr 'brass' field guns and 1 24-pr howitzer	48
	10 gun ammunition wagons	60
	2 howitzer ammunition wagons	12

1 spare gun carriage	6
1 forge wagon	6
2 store wagons	12
1 store cart	2

Riding	20
Baggage	8
Spare	18
Total	192

(*b*) *9-pr Field Battery*

As for a 12-pr battery, but with only 7 ammunition wagons, 1 store wagon, 19 riding horses, and 15 spare horses.

(*c*) *6-pr Light Field Battery*

As for a 12-pr battery, but with a 12-pr howitzer, 6 ammunition wagons, 1 store wagon, 18 riding horses, and 13 spare horses.[11]

All these field pieces were 'brass'. The guns fired round shot, common shell, shrapnel, and case shot; whilst the howitzers fired common shell, shrapnel, case shot, and carcasses.

By the time of the Crimean War there were in each battery four guns and two howitzers. The 6-pr gun was associated with the 12-pr howitzer and the 9-pr gun with the 24-pr howitzer. The object was to enable any battery to suit its fire to the ground and the circumstances of the time to produce the effect that its role required; and this equipment policy was maintained until the end of the smooth-bore period. By January, 1854 all horse artillery had been armed with the heavier pair of these equipments; but a reversion was made on the outbreak of war, and the first two troops to embark for the theatre of war did so with the lighter pieces. All the field batteries, however, went out with the heavier equipment.[12]

Of the heavy smooth-bore pieces, the 68-pr gun was the most powerful in the Service, and the latest type in use was introduced by Colonel Dundas, R.A., in about 1840.

79

But it had a respectable ancestry, for it was descended from the old 'cannon royal' with a similar calibre. It replaced the 56-pr which had been produced some two years earlier. Both were normally used for coast defence, but some of the 68-prs (also an armament of H.M. Ships) were used on shore, manned by naval ratings, for the reduction of Sevastopol. The 42-pr gun appeared occasionally in the siege train during the Peninsular War, but was not afterwards included in a field army. The 32-pr gun was more generally used, both on land and at sea, than any other heavy gun and was a very effective piece at long ranges. It was equivalent to the old demi-cannon. There were a number of different types of 24-pr gun, but the 50-cwt version used during the Napoleonic Wars proved to be the most useful of all British ordnance for breaching walls and other defensive works. With round shot it could pierce twelve feet of rammed earth. It was used at all the Peninsular and Mutiny sieges, and Miller says (*Equipment of Artillery*), 'It may be drawn by cattle of the country in such teams as the state of the roads renders necessary. A bullock pole is provided to be used with common yokes for oxen.' In sieges the 18-pr gun was used principally to neutralize the opposing artillery whereas the batteries of 24-prs or other heavy pieces were being established for their breaching task, and it was very successful in this role at the siege of Delhi. In the Sikh Wars the 18-pr was used most effectively against an enemy field artillery which was far superior to that of the Bengal Army and which would otherwise have caused heavy casualties amongst the attacking infantry. But it was at Inkerman, in the Crimean War, that this piece acquired immortal fame, when two 18-pr guns destroyed the whole of the great Russian battery covering the enemy attack and played a decisive part in the Russian defeat.

The 10-in. and 8-in. iron howitzers were introduced by General Miller in about 1820; previous pieces of this type having been made of 'brass'. The 10-in. was primarily a siege weapon and it was at one time much used for ricochet fire. The 8-in. was a very popular equipment, and from the latter part of the Peninsular War onwards it was often used as heavy artillery in the field instead of the 18-pr gun.[18]

Large mortars were made of 'brass' up till 1780, when a start was made in casting them in iron. For these large mortars there was an octagonal bed, above which was a wooden 'deck', mounted on a pivot so that the piece could be traversed. The heavy sea service 13-in. mortar was employed during the latter part of the siege of Sevastopol; but the land service 10-in. and 8-in. mortars were the pieces normally included in the siege train. The sling wagon would carry the land service mortar already mounted on its bed. The block carriage, or platform wagon, would take two 8-in. mortars with their beds. The sling wagon and sling cart (a two-wheel version of the sling wagon) were also used for the 8-in. mortar – the former to carry mortar and bed together and the latter separately. The small 5½-in. royal and 4⅖-in. coehorn mortars could each be carried about by a team of four men and could be used anywhere without the necessity of laying a bed.

One peculiar difference between pieces was that those firing shot generally had three muzzle mouldings, whereas those used for shell had only two.[14] Guns at the end of the smooth-bore period were 14 calibres and upwards in length, carronades were 7 calibres, howitzers 4–10 calibres, and mortars 3–4.[15] There were some minor differences between pieces intended for sea service (SS) and those for land service (LS). SS ordnance, for instance, were always furnished with a breeching loop (i.e., a loop on top of the button). As late as 1887 there were 58 different types and varieties of smooth-bore ordnance of cast iron and 'brass', listed as to be retained in the Service.[16]

The deficiencies of 'brass' ordnance became apparent during the sieges of the Peninsular War. At the beginning siege trains were composed entirely of 'brass' ordnance. But the heat of rapid fire so softened the 'brass' that the muzzle began to droop. As a result the limit for 'brass' equipments was set at 120 rounds in twenty-four hours; iron guns could fire three times the number in the same period. This so hampered the rate of fire that from 1811

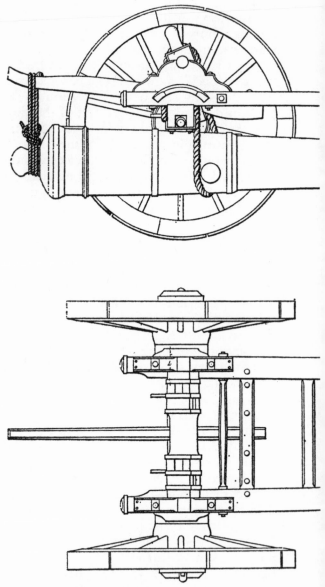

Sling wagon for 24-pr gun.

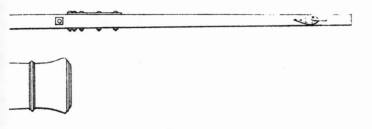

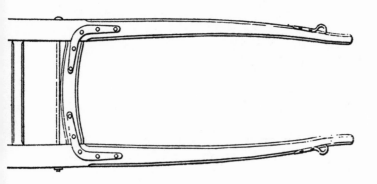

onwards only iron guns were allowed in British siege trains. In 1816 all 'brass' pieces were abolished except for the 12-pr gun and the 32-pr howitzer.

General Paixhans, a famous officer of the French Artillery, was responsible for the design of a gun intended to fire shell. As compared with contemporary ordnance, his gun had a distinctive shape, for he had discarded all useless metal and it had a straight muzzle, smooth lines, and none of the customary rings and other adornments. Other countries subsequently made shell guns and copied this pattern, which became known as the Paixhans gun. It was demonstrated impressively at Brest, firing at a ship, in 1821 and 1824. In 1837 the French decided on a gradual change to shell guns; and in 1839 the British decided to follow suit, though initially the shell was only introduced as an auxiliary to the solid round shot.[17]

In pursuance of their decision, the French determined that a 12-pr should constitute the only calibre for their field artillery; but as their existing equipment of this nature was very heavy, the Emperor Napoleon III, in 1853, designed a 12-pr gun/howitzer to fire common shell, and this was the standard French field equipment in the Crimean War. This piece was introduced into the United States Army in 1857 as the single armament for the field artillery and was the most popular field equipment in the American Civil War.[18]

Gun Carriages

A disadvantage of the bracket trail was that its weight and length combined to move the point of balance so far to the rear that moving the trail to traverse the piece was very hard work. In 1792 General Sir William Congreve introduced a much lighter and shorter 'block' trail which was a solid piece of narrow rectangular section. This moved the centre of gravity forward and traversing became a much easier task. The block trail was used from the Peninsular War onwards for 6-pr and 9-pr guns, and for

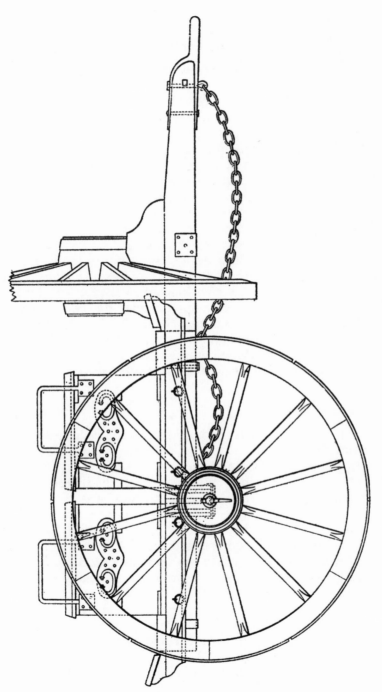

Ammunition wagon, 1852.

12-pr and 24-pr howitzers when these latter were introduced in 1822.

Bracket trails were retained for heavy guns until the 1850s, as were also the old-fashioned limbers, consisting merely of axletree, wheels, and pintle. The limber wheels were smaller than the gun wheels so that the trail should be reasonably low. On the gun carriage there were separate 'travelling' trunnion holes, about two feet to the rear of the normal ones; and for movement the piece was lifted into the travelling holes to distribute the weight better for movement. During the nineteenth century the quoins used to elevate the piece were replaced by an elevating screw.

The carriages of the heavy 8-in. and 10-in. howitzers were quite different from all the others. The piece was supported in front by two long brackets, and at the rear by a short trail called a perch, under the end of which were two small truck wheels. On either side of the brackets and hinged to them at one end were two long brake levers. The free ends of these could be lashed down to bear on the naves of the truck wheels and thus control the recoil.[19]

Mountain Artillery

Pack artillery was first used by British forces in the Bhutan expedition, on the north-east frontier of India, in 1772; the equipments being carried by coolies. In 1813, when Wellington's advance through Spain had reached the Pyrenees, the first Mountain Battery was formed by Lieutenant W. L. Robe, R.A. It was equipped with six 3-pr guns, each of which weighed 252 lb and could be carried on one mule. The gun carriage was dismantled and loaded on to two other mules. In 1816 a mountain train of $4\frac{2}{5}$-in. brass howitzers and mortars proved most useful in Nepal. (The mortars, which were cast in 1738, were probably the oldest mountain guns in service in any army.) The $4\frac{2}{5}$-in. howitzer was also the equipment of the Pack Artillery formed as part of the British force dispatched to Spain in 1836 to fight the Carlists.[20]

The first permanent 'mountain train battery' (the Hazara

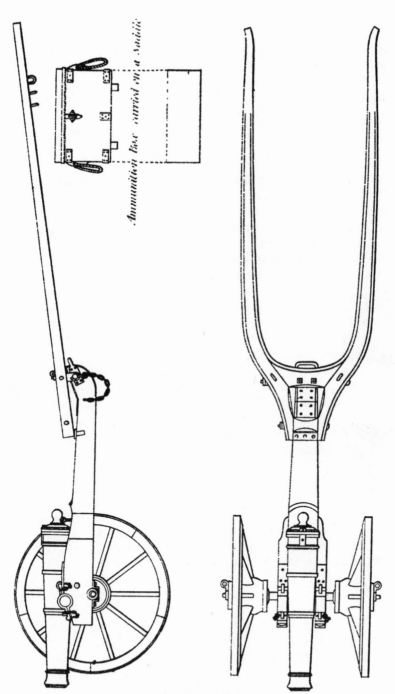

Mountain artillery, 1845.

Mountain Train) was formed in India in 1850 and equipped with six 3-pr brass guns. These were split up into mule loads, each load consisting of a gun, trail, wheels and two pairs of ammunition boxes; any extra ammunition being carried on hired mules.[21]

The Vent

The vent was a very small but most important feature of muzzle-loading ordnance. It was a small hole at the breech end of the piece which was not quite vertical, inclining at an angle of 101° to the forward axis of the piece. The internal diameter had to be very small because so much gas would otherwise escape through it, and it was fixed at $\frac{1}{5}$ in. Even so, the gas which did escape gradually burnt away and enlarged the vent till, during the Peninsular War, the vent often became so large that the piece was unserviceable. To avoid this, the vent was lined by inserting copper, which was less liable to burning than other metals. This process was known as 'bouching', and from about 1820 all new pieces were issued already bouched.[22]

A muzzle-loading piece could be put very effectively out of action by 'spiking' it. The spike used was 4 in. long, .27 in. in diameter at the head, and .1 in. in diameter at the point. It was driven as far into the vent as it would go and then broken off close to the piece.[23]

Shrapnel

In 1784, Lieutenant Henry Shrapnel R.A. invented what he called a 'Spherical Case Shot', which was a hollow iron sphere containing a number of bullets, a bursting charge, and a fuse to fire the charge. The invention was demonstrated before the General Officer Commanding Gibraltar in 1797[24] and was finally approved for Service use in 1803–4.[25]

The first recorded use of the new projectile in action was at the battle of Rolica.[26] In the original design the bursting charge was in direct contact with the bullets, and there were a number of premature explosions caused by the friction of the bullets against each other. To remedy this the powder charge of the 9-pr gun was diminished by half a pound. Captain (later Major-General) E. M. Boxer R.A. suggested an improved design in which the charge was enclosed in a tube to keep it from contact with the bullets; and in 1840 he improved this again by using a diaphragm to separate charge and bullets.[27] Boxer's first pattern was officially approved in 1854 and his later design in 1856.[28]

Laying and Firing

The laying and firing of a smooth-bore muzzle-loader was quite a complex business. To lay the piece it first had to be traversed and then aligned on its target in the vertical plane. The process was completed by elevating the piece to allow for the distance the projectile would fall during its flight. Traversing was effected by moving the trail; aligning was carried out by looking along the 'line of

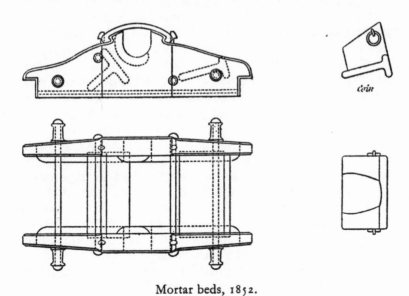

Mortar beds, 1852.

metal'; i.e., from the top of the base ring to the top of the muzzle. This line, however, was not parallel to the axis of the bore, and the difference had to be allowed for. On some pieces, especially howitzers, a foresight, known as a 'dispart patch', was cast on top of the muzzle ('dispart' was the difference between the half diameter of the piece as measured at the base ring and the muzzle).

Mortars could only be traversed because they fired at a fixed elevation of 45° and the range was adjusted by altering the charge. A line was drawn with chalk along the centre of the piece from base to muzzle. The layer held a plumb line aligned on the target and the mortar was traversed till the chalk line coincided with the plumb line.

The next procedure was to load the piece. The loader first placed a charge in the muzzle. (If a round had already been fired, the spongeman dipped his sponge, on the end of the long sponge staff, into a bucket of water carried on the gun carriage and pushed right down the bore to extinguish any smouldering remnants there might be from the previous charge.) Reversing the sponge staff, the spongeman rammed the charge home; at the same time the ventsman 'served the vent' (i.e., put his thumb over it) to prevent a rush of air which, if the sponging had still left any burning fragments, might cause a premature explosion of the charge. The ventsman then pushed the 'pricker' (which resembled a long skewer with a handle on one end) down the vent to pierce the serge wrapping of the cartridge. The loader having put the projectile in the muzzle, this too was rammed home by the spongeman. The firer inserted in the vent a goose quill filled with a composition of mealed powder and spirits of wine and, from the slow match of the linstock, lit his portfire (a stiff paper tube 16½ in. long filled with a composition that burned at the rate of one inch per minute). On the order to fire he applied the portfire to the goose quill and then cut off the burning portion of the former with the portfire cutter attached to the trail of the gun carriage.[29]

By 1853 a friction tube had been introduced in place of the goose quill. A spark was produced to ignite the charge

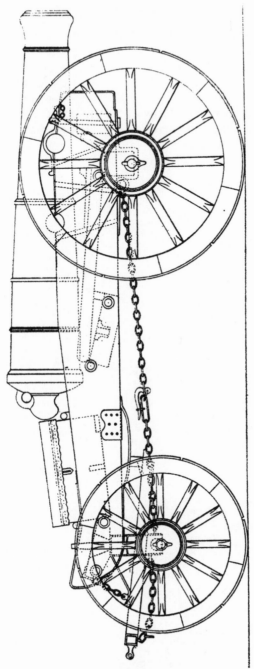

18-pr gun and carriage.

by rotating a roughened metal bar, inserted in the top of a tube filled with a composition, by a pull on a lanyard.[30]

A major change in the direction of the Royal Artillery was made as a result of the breakdown in administration in the Crimea. The Board of Ordnance was abolished and Artillery and Engineers now came under the direct control of the War Office and the Commander-in-Chief.

CHAPTER VIII

The Advent of Rifled Ordnance

IN his *Abridged Treatise on the Construction and Manufacture of Ordnance in the British Service* of 1877 (an official manual), Captain J. F. Owen says:

> When rifled small arms were generally adopted about 1855, it became necessary to introduce rifled ordnance in order that artillery might still remain the most powerful arm in the field; for infantry skirmishers, armed with long range rifles could do much damage to the men and horses of smoothbore batteries, while they themselves were out of dangerous range of the guns.

Although rifled ordnance was first used in the British Army during the Crimean War, there had been a number of experiments in that direction since the beginning of the seventeenth century. That they had not been successful was due to the lack of sufficient skill in working metals and of the machinery for boring and rifling with the necessary accuracy.

The first British rifled ordnance consisted of a few old 68-pr and 8-in. cast iron guns, which were made oval and twisted in the bore and so converted to pieces rifled on the Lancaster principle (devised by Charles William Lancaster, 1820–78). They were used at the siege of Sevastopol, but their shooting was irregular and the projectiles often jammed in the bore, and ruptured the chase.[1]

The battle of Inkerman led to investigations into rifled ordnance by a remarkable man. William George Armstrong was a Northumbrian, born at Newcastle-on-Tyne in 1810

93

and son of a corn merchant who became Mayor of the City. Although he studied law and became partner in a legal firm, his genius as an inventor was demonstrated at an early age. In 1839 he constructed a 'water-pressure wheel' and in about 1844 a hydro-electric machine. In 1845 he became secretary to a water company, the following year he patented a hydraulic crane, and a year later he was the first manager of Elswick-on-Tyne Engineering Works. The Crimean War saw his genius applied to military inventions. In 1854 he designed a submarine mine, and then the battle of Inkerman drew his attention to the possibilities of rifled ordnance. As stated in the last chapter, the result of the battle of Inkerman was decided by two 18-pr guns; but the business of dragging these heavy pieces uphill to a close enough range showed the need for an accurate and light field gun with a long range. Armstrong saw that the solution lay in a rifled weapon, made of forged, instead of cast, iron. In December, 1854, therefore, he suggested to the Duke of Newcastle, Secretary of State for War, the need for a rifled field-gun firing an elongated lead projectile. The Duke of Newcastle was interested and Armstrong was given an order for six guns.

Armstrong was not alone in seeing the value of rifled ordnance, in the light of Crimean experience. Napoleon III, an acknowledged authority on artillery, had become convinced that smooth-bore pieces were inadequate for the artillery task. After the war he had some 'brass' pieces rifled and sent to Algeria, where in active operations they soon proved their great superiority over the equivalent smooth-bore field guns. All the 'brass' field pieces in the French Army were accordingly rifled on a system proposed by M. Treuille de Beaulieu in 1842. The de Beaulieu rifling consisted of six shallow and rounded grooves in which ran zinc studs formed in two successive bands on a cylindrical projectile. This form of projectile showed another advantage of rifling in that the spin imparted to it ensured that the point would hit the target or ground first; and it was therefore possible to use fuses which could be actuated on contact. At the battles of Magenta and Solferino in 1859 the French 'brass' rifled guns were used with devastating effect.[2]

After two years of development Armstrong's first gun was ready, and was tested with impressive results. It had a steel barrel with polygroove rifling (i.e., a large number of shallow grooves) with a uniform twist from breech to muzzle. Externally the barrel was supported by wrought iron hoops, made by winding a long bar when hot round a mandril so that it formed a cylinder with the appearance of a tightly coiled helical spring. This cylinder was then welded into a solid wall and machined to the right dimensions for fitting either directly over the steel barrel, or as the outer layer over a similar cylinder. The cylinders were then expanded by heat, fitted respectively over the barrel and each other, and allowed to cool and shrink. The forged steel barrel was not a success, owing to the lack of knowledge of steel at the time, and Armstrong used wrought iron for the barrels of his subsequent pieces.[3] The Armstrong gun was a breech-loader, with a plug, or block, placed in a slot in the breech of the gun and screwed up against the end of the barrel by a hollow screw, through which the gun was loaded.[4] The projectile was iron coated with lead, in which soft metal the rifling engaged. The piece was mounted on a carriage which allowed the gun to recoil up an inclined slide and return by gravity.[5] Its accuracy as compared with a smooth-bore piece of similar calibre was astounding. A 12-pr rifled breech-loader could put one shot out of two in a space measuring 25 yards by $\frac{1}{3}$ yard, whereas a smooth-bore gun needed 121.7 yards by 25.8 yards.[6]

Besides Armstrong's, another system soon appeared. Between 1854 and 1857 Mr J. Whitworth manufactured some breech-loading guns with his own hexagonal rifling. On test these achieved a standard of range and accuracy higher than any previously recorded.[7] Joseph Whitworth (1803–1887) had worked first as a mechanic in Manchester and London and had set up as a tool-maker in Manchester in 1833. There he achieved fame with his measuring machine for his system of standard measures and gauges.

In 1858 Parliament appointed a committee to examine and report on the merits of the different systems of rifling ordnance. The committee soon came to the conclusion that only the Armstrong and Whitworth systems were worthy of consideration. However, the tests on the Armstrong gun had been so satisfactory that the committee reported in favour of it and it was forthwith adopted for field service. In January 1859 the Government entered into a contract with Armstrong's newly established Elswick Ordnance Company and also began manufacturing rifled breech-loaders (R.B.L.) at the Royal Arsenal, Woolwich. Armstrong handed over the rights of his gun to the nation; and for his invention and the gift he was knighted and appointed Engineer of Rifled Ordnance to the War Department. Soon afterwards he was appointed Superintendent of the Royal Gun Factory at Woolwich.[8] Before long, horse and field artillery were equipped, respectively, with Armstrong 9-pr and 12-pr rifled guns, and heavier pieces were being made for siege and garrison artillery and for the Fleet.[9]

Meanwhile, by 1860 Whitworth had developed his invention and had produced pieces made of wrought iron cylindrical tubes forced over one another by hydraulic pressure, as compared with the Armstrong system of heating and shrinking. One of his pieces, a 3-pr gun, fired a shot over the then record distance of 9,688 yards.[10]

Armstrong's breech-loading system was complex and rather in advance of its time, and there were a number of accidents due to various mechanical weaknesses. Nevertheless, the Armstrong rifled breech-loading ordnance was very successful, both on land and sea, in various minor campaigns.[11]

In 1864, owing to the objections raised against the breech-loaders, Armstrong introduced 40-pr and 64-pr guns with wedge instead of screw breech actions and 64-pr muzzle-loading guns. In addition he submitted a proposal for muzzle-loading field guns. Whitworth also proposed a new muzzle-loading system. A special committee was appointed to carry out trials with 12-pr and 70-pr

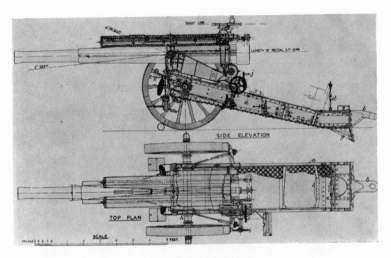

9.   Mark I 60–pr BL gun.

10.   13–pr gun team of the RHA at Inexent, 1918.

11. The 13-pr AA gun. Lievin, 1918.

rifled guns, constructed respectively as Armstrong breech-loaders, Armstrong muzzle-loaders, and Whitworth muzzle-loaders. The 12-pr had been chosen for a possible field gun and the 70-pr as an equipment suited for siege, garrison, and naval broadside. Whitworth's guns were rifled on his hexagonal system. Armstrong's 12-pr breech-loaders had his breech screw whereas the 70-prs had the wedge system. His muzzle-loaders were built up on the same principle as his breech-loading pieces.

On test all guns were very good; but the Committee reported in 1865 that the muzzle-loaders were far superior to the breech-loaders in range, accuracy, ease of working, endurance, and cost.[12]

But though it was clear that the future lay with rifled ordnance, both Services possessed thousands of cast iron smooth-bore equipments, and to scrap them all would have been extremely expensive. Attempts at rifling smooth-bore pieces in the American Civil War had not been very successful. In 1862 many of them were rifled and then reinforced by shrinking on iron hoops; but this sytem did not provide sufficient strength and many pieces burst. In 1863 a scheme proposed by Major (afterwards Sir William) Palliser R.A. was adopted. Tubes of coiled wrought iron were fitted inside the cast iron barrels and rifled, so that the converted piece consisted of a wrought iron inner tube supported by a cast iron jacket.[13]

In 1868 the field artillery in India still consisted almost entirely of smooth-bore pieces. Re-equipment from the United Kingdom was difficult because at that date the manufacture of breech-loading ordnance had ceased and the construction of rifled muzzle-loaders (R.M.L.) had not begun. It was decided that in any case India should be able to make the field guns required to equip the forces at the disposal of the Indian Government; but, on the other hand, it would be expensive to provide and send out the plant necessary to make steel and wrought iron rifled ordnance. The solution ultimately reached was that the Army in India should have bronze rifled muzzle-loaders. Trials at Woolwich with a 9-pr R.M.L. bronze gun, cast at Woolwich,

97

gave such good results that this pattern was selected for the field artillery in India, and some were even issued to batteries in the United Kingdom. So many defects, however, developed in these bronze guns that all those in the United Kingdom had been withdrawn by 1877.[14]

On 25 July, 1870 a special committee of officers of the Royal Artillery under Major-General Sir John St George was appointed to carry out trials with the Indian pattern 9-pr bronze R.M.L. gun and with the 9-pr and 12-pr wrought iron and steel R.B.L. guns of the home service. The report of the Committee is of great interest and its major points were as follows:

> The Committee have no hesitation in giving preference to the M.L. gun, both in respect to simplicity and facility of repair.
>
> If, as regards the question of endurance, the Committee are called upon to select between a structure of wrought iron and steel and one of bronze, as tried by them, they unhesitatingly pronounce in favour of the former, whether the gun be a muzzle-loader or a breech-loader.
>
> It may be urged that considerations of manufacture favour the adoption of bronze for M.L. guns for India, but the same cannot be accepted as applying to this country, and taking weight for weight it is impossible to deny that far greater endurance will be attained by the present mode of construction than by the use of bronze.
>
> The service B.L. system, owing to the absence of windage, necessitates the employment of a mechanical arrangement for lighting the time fuze. This is effected by the employment of a detonator, which has proved to be highly sensitive to climatic influences. The M.L. gun on the other hand has the advantage of being able to use an ordinary wood time fuze, which experience has proved to be little or not at all affected by climate.
>
> As regards cartridges, the B.L. guns have the great disadvantage of requiring the use of lubricators.
>
> In respect to other stores, such as percussion fuzes and

projectiles, the Committee believe that, whether for breech or M.L. guns, there will be found little or no difference between them so far as regards their capability to bear the tests of travelling or climate.

Judging from the results of the practice at Aldershot, the 9-pr. M.L. and the 12-pr. B.L. guns appear, in respect of shooting to be much on a par; the former being superior in point of shrapnel shell with time fuzes, the latter in point of segment shell with percussion fuzes; the 9-pr. B.L. gun being inferior to both.

The advantages of simplicity, facility of repair, ease of working, rapidity of fire, original cost, and cost of maintenance, are in favour of a M.L. gun, and the Committee consider that these qualifications outweigh the important advantage of the superior amount of cover given to the detachments when entrenched and in the open, which a B.L. gun affords, and are therefore of opinion that on the whole, a M.L. gun is the more efficient for war purposes; but they recommend that, if adopted for home service, they may be made of wrought iron with steel tubes.

No R.B.L. ordnance was made for many years after 1864, and the manufacture of R.M.L. steel and wrought iron ordnance began in 1871. They were built on the Armstrong principle of an inner barrel of steel and jackets of coiled wrought iron.

Armstrong resigned his appointment at Woolwich on account of the official preference for muzzle-loaders and carried out research into and the development of improved breech-loading ordnance. Whitworth, in his turn, was created a baronet in 1869.

An earlier Committee of Royal Artillery officers under Major-General C. Dickson, V.C., C.B., had assembled at Dartmoor in June, 1869 to determine the best projectiles and fuses for B.L. guns, and they took occasion when writing their report to recommend a field howitzer which could be a muzzle-loader suitable for high-angle firing and capable of throwing a common shell with a large bursting charge. As a result of this recommendation a high-angle R.M.L. 16-pr 12-cwt gun was approved as the equipment for half of the field batteries. It was then the most powerful field piece in Europe. The other half of the field batteries and the horse artillery were equipped with R.M.L. 9-pr 8-cwt guns; but in 1874 the horse artillery received the lighter R.M.L. 9-pr 6-cwt guns, and these later replaced the 8-cwt guns in field batteries.[15]

The important part played by the Prussian field guns in the war of 1870 impressed British military opinion, and the War Office decided to test the new British R.M.L.s against the Prussian R.B.L.s. In 1872, therefore, trials were carried out at Shoeburyness between a 9-pr field gun of each type. The committee in charge reported that the British gun was superior to the German in range, rapidity of fire, and ease of manipulation, though the German gun was slightly the more accurate. They added that throughout the trials the breech mechanism of the German gun required much attention. It was noted also that this complicated breech action gave so much trouble during the war of 1870 that at one time 200 guns were unserviceable. The Germans had since introduced a new gun with an improved and simpler mechanism.[16]

Also in 1872, the French tried the Woolwich 9-pr R.M.L. at Bourges against their own and other nations' field guns. The official report said: 'The Woolwich wrought iron 9-pr gun gave results which are not inferior to those of any gun actually in service in Europe.'[17]

In 1871 a 25-pr 18-cwt R.M.L. was approved as 'a gun of position' and introduced three years later. Early in 1877 a 13-pr 8-cwt R.M.L. gun was designed at the Royal Gun Factory to replace the 9-pr in horse and field artillery.[18] It proved unpopular, however, owing to its excessive recoil, and issue was stopped before all batteries had been re-equipped with it.[19]

Rearmament for siege artillery after 1871 consisted of 25-pr, 40-pr, and 64-pr R.M.L.s, replacing, respectively, 20-pr, 40-pr, and 7-in. R.B.L.s. For high-angle fire, smooth-bore mortars remained in use long after other S.B. ordnance had been abolished. Rifling of mortars was not a success, because, firing as they did at fixed angles, rifling did not improve their accuracy. In 1872, however, it was decided that mortars should be replaced in the siege train by 8-in. 46-cwt rifled howitzers, and a number were made during the next two years. The success of this piece led to the introduction of another and lighter 6.3-in. howitzer of 18 cwt. These two howitzers eventually replaced all the old smooth-bore mortars.[20]

Mountain Artillery

Mountain artillery has a peculiar fascination, due, perhaps, to the problems which have to be surmounted to produce a piece which can be rapidly dismantled and assembled, and which is divisible into loads within the carrying capacity of a mule. There is an added attraction in the difficult country for which the equipment has been designed.

A 6-pr R.B.L. was in 1858 recommended for mountain service, but it proved too heavy for a mule and had to be shortened. Six of them were used in the Bhutan campaign of 1864 and did well, but they were not as good as the $5\frac{1}{2}$-in. mortars which they were intended to replace. Even when shortened, they were too heavy for mules to carry and they were generally drawn by mules on their carriages. Sometimes in India, however, they were carried on elephants.[21]

During this Bhutan campaign the home authorities were asked for some light steel rifled guns. None, however, could be made available in time, so some of the old 3-pr 2.9-in. bronze guns were bored out to three inches, rifled, and lightened to two cwt. They arrived too late for use in the war, and, as they were too heavy for mules, they had to be lightened still further. They were used on the Red River expedition of 1870, but were not very successful, and manufacture ceased in 1873. They were either the last, or nearly the last, bronze guns in British service.[22]

The request for the Bhutan expedition had, in the meantime, resulted in the construction in 1865 of a 7-pr

3-in. rifled muzzle-loading gun weighing 150 lb, which was used effectively in the Abyssinian war of 1868. In 1873 a modified version was adopted, with the bore lengthened from 24 in. to 36 in. and the weight increased to 200 lb.[23]

In 1877 Colonel Le Mesurier R.A. proposed a gun on an entirely new system. It was to consist of two parts, each one mule-load, and joined by a loose trunnion ring. This was attached to the muzzle portion and screwed on to the breech portion. Newspaper reporters called this piece the 'screw-gun', and as such it was immortalized by Rudyard Kipling. The length of the bore was 66½ in., its diameter was 2½ in., and the projectile weighed 7 lb. This gun was adopted as the standard equipment for mountain artillery in 1879, and it was then, and for many years afterwards, the best mountain gun in the world. It was originally called the '7-pr R.M.L. jointed steel gun of 400 lb', but this mouthful was soon changed to '2.5-in. R.M.L.'[24] It was first used in action in 1879 on the North-West Frontier of India by a British mountain battery, 1/8 R.A. The whole equipment constituted a load for five mules: one for each portion of the piece, a third for the carriage, a fourth for the wheels, and a fifth for the remainder (axletree, elevating gear, cascabel block, hammer, and stores).[25] This little gun remained the armament of mountain artillery until after the Boer War.[26]

Construction of Rifled Ordnance

The deficiencies of bronze (or 'brass') as a material for ordnance have already been mentioned. Phosphor bronze was tried, but without marked improvement. In 1872 the French Government carried out trials at Bourges with two 4-prs of ordinary bronze and two made of phosphor bronze, but only slight improvement was shown by the latter. In Austria the attention of General von Uchatius, Director of the Vienna Gun Foundry, was drawn to a fragment of bronze cast under pressure, which the Archduke William had brought from Russia. He found it so superior to ordinary bronze that he directed research into its manufacture. The result was that the Vienna Foundry began casting bronze in an iron mould, chilling the interior of the mass by means of a core of solid copper. This bronze, which was an ordinary alloy containing 8 per cent tin, could be forged cold and it possessed so many of the qualities of steel that it was sometimes called 'bronze steel'. Experimental guns made of it were so successful that all subsequent Austrian rifled field guns were made of it.

Cast iron was a suitable metal for smooth-bore ordnance, for it could easily be formed in the required shape and its comparative weakness did not matter because the pressure in the bore was not very great. But it was not normally strong enough by itself for rifled ordnance; though in Sweden and Denmark, where exceptionally good cast iron was produced, rifled field guns were still being made of cast iron in 1877.[27]

Wrought iron was used for the barrels of early rifled ordnance, as already mentioned, but the material was soft and liable to flaws, so that by 1877 no British heavy gun barrels were made of it. Steel was an alloy made by casting iron while in a fluid state into a malleable ingot. But at this period its brittleness and unknown qualities militated against its use as the sole material in the manufacture of ordnance, except for the smaller pieces (such as mountain guns). Nevertheless, it was admirable for the construction of an inner barrel surrounded by a wrought iron jacket. In 1877 the barrel was made of mild steel, whereas good ductile wrought iron was used for the rest of the piece. This type of manufacture proved very satisfactory, and of the thousands of pieces of ordnance built on this principle none ever burst in use. Foreign guns built on different principles could show nothing like this record. For instance, R.M.L.s made on the Parrott system of cast iron strengthened by wrought iron superimposed rings were used extensively in the American Civil War and proved most unreliable. The Report on Ordnance to the United States Senate of 25 February, 1869 stated: 'In the attack on Fort Fisher all the Parrott guns in the fleet burst. By the bursting of five of these guns at the first bombardment 45 persons were killed and wounded, while only 11 were killed or wounded by projectiles from the enemy's guns during the attack.' In France, in 1875, a 170-mm. gun of cast iron with steel barrel and steel hoops burst. A number of Krupp steel guns burst during the 1860s, six, for instance, during the Austro-Prussian War of 1866.[28]

Casting was undoubtedly the simplest form of manufacture; but it could not be given strength above a certain limit, beyond which any additional thickness of the walls added little to the ability of the piece to withstand pressure. Shortly after the projectile begins to move, the pressure inside the barrel lessens, and goes on decreasing as the projectile approaches the muzzle. The piece, therefore, was made stronger about the powder chamber than towards the muzzle; and this relative difference in strength was necessarily greater in rifled than in smooth-bore ordnance because the pressure in the former was much greater and the projectile did not move so soon.[29]

British R.B.L.s and the earliest of the heavy R.M.L.s were built on the Armstrong principle in which the inner steel barrel was jacketed by a number of thin wrought iron coils shrunk over one another in concentric layers, and which incorporated a solid forged breech piece.[30]

Mr Fraser, Deputy Assistant Superintendent of the Royal Gun Foundry, proposed in 1865 the use of larger wrought iron coils so that a forged breech piece would be unnecessary. This proposal was adopted for some of the larger types of ordnance.

The screw breech action has already been mentioned. The breech block, which was pressed against the end of the barrel, was called the 'vent piece' because through it passed the vent channel. These vent pieces dropped into place from above, and were lifted away from the piece after each round.[31] They were extremely heavy and were suitable only for the lighter artillery; this was the reason why Armstrong invented his wedge, or side-closing arrangement, for the larger pieces.

Ammunition

In 1878 Armstrong's Elswick firm carried out experiments with slow-burning powders and long guns. To obtain slow-burning, a large grained powder was used, and this, used in a long gun, gave very satisfactory ballistic results. Powder with a small grain burnt so quickly that there was no advantage in having a long barrel to make better use of its potential power. With a larger grain, the time taken to ignite the whole was increased and the pressure behind the projectile was maintained for a longer time and a long barrel became an advantage. This large-grained powder was termed 'pebble powder'. It had not been long in use before it was found that still better results could be obtained by igniting the grains from the centre instead of from the outside. Grains, therefore, were made in the shape of a hexagonal prism with a hole in the middle, and in a powder charge the prisms were packed together so that the central holes coincided. This prismatic powder remained in use until superseded by smokeless explosives.[32]

Powder was packed into cartridges made with serge and covered with worsted until about 1878, when it was replaced by a material known as 'silk-cloth'. Cartridges were packed in either metal-lined cases, barrels, or ammunition boxes. The metal-lined cases were in three sizes, known respectively as 'whole' (containing 120 lb of powder), 'half' (60 lb), and 'quarter' (30 lb). They were made of wood and the lining was copper, and were primarily for use in damp magazines and during siege operations. The barrels were either whole or half (i.e., 120 lb or 60 lb) and were used in dry magazines. Ammunition boxes were issued to field batteries at home and were marked with the nature and number of the cartridges.[33]

It was really the French design of projectile which permitted the replacement of rifled breech-loaders by rifled muzzle-loaders. Armstrong's lead-coated shell could not be pushed down the muzzle and was therefore limited to breech-loaders. The French projectiles had studs in the side which fitted in the few shallow grooves of the rifling (as compared with the polygroove rifling of the Armstrong breech-loaders). This type of projectile could be easily loaded through the muzzle. With the studded sides, however, there was some windage and this resulted in erosion of the bore. To remedy this a papier-mâché cup, called a gas check, was inserted between the base of the shell and the cartridge. In 1878 the papier-mâché gas check was superseded by one of copper. The next step was to fix the gas check to the base of the shell, and, since the tight-fitting copper made the shell rotate, the studs could be dispensed with. This led to the copper driving band being incorporated as a component part of the shell and, as there were no studs, a reversion could be made to the earlier polygroove rifling.[34]

Ordnance in use in 1877

It is perhaps worth listing the rifled ordnance in use in the British Services just before the re-introduction of the rifled breech-loader. These comprised the original R.B.L.s, their successors the R.M.L.s, and R.M.L.s converted from smooth-bore muzzle-loaders.

(a) Rifled Breech-loading Ordnance

A 7-in. 72-cwt gun was recommended to the Navy in 1859 as a broadside or pivot gun in replacement of the old smooth-bore 68-pr. This 7-in. piece, however, was too light and produced a recoil which was too great for its use on board ship; only six, therefore, were made. A heavier 7-in. of 82 cwt was produced in 1861, to replace the 72-cwt, for both land and sea service. This too was an unsatisfactory equipment, for it was not powerful enough to penetrate armour. In due course it was replaced by a R.M.L.

In the same year as the 7-in. (or 100-pr), a 40-pr 32-cwt was recommended as a naval broadside or pivot gun, and it was later adopted for land service use for 'batteries of position', siege work, and garrisons. It was sometimes referred to later as the O.P. (old pattern) 40-pr. A heavier

40-pr of 35 cwt was introduced in 1860, known as the 'G' pattern. It had a longer and stronger breech piece. It was constructed, rather oddly, as a precaution in case any weaknesses should develop in the 32-cwt, though no symptoms of weakness had as yet appeared – nor ever did!

Three 20-pr guns were introduced in 1859, one for land service (L.S.) and the other two for sea service (S.S.). The L.S. 16-cwt equipment was intended as a light gun 'of position'. It was at first a 25-pr, but the projectile was lightened and it was used in heavy field batteries of reserve. Of the sea service pieces, one was a 15-cwt broadside gun for sloops, which had insufficient beam to work the 16-cwt. It was 2½ feet shorter than the L.S. equipment. The other S.S. piece was a 13-cwt 'pinnace' gun for boat and field marine use. In 1877 it was used on the upper decks of ironclad ships for action at close quarters to repel boarders and for firing at torpedo boats.

A 12-pr 8-cwt gun appeared in 1858 for land service use as a field gun, and it was subsequently adopted by the Navy, reduced in length by 12 ins, as a boat or field marine gun. In 1863 a modified version was made for use by both Services. It was the same calibre as the 9-pr, and could use 9-pr ammunition, though the reverse did not apply.

The 9-pr, referred to above, was a 6-cwt gun of 1862, primarily for the Horse Artillery, but the Navy also acquired some as boat or field marine guns.

The 6-pr 3-cwt gun, mentioned before, intended for mountain service but found too heavy to be carried by a mule, was supplied in some numbers to Colonial batteries, and the Navy got some to add to their comprehensive boat and field marine armoury.

(b) Rifled Muzzle-loading Ordnance

To start at the other end of the scale for a change: the smallest R.M.L. gun was the 7-pr. The first model was the Mark I bronze piece, which was produced in a hurry to meet the demands of the Indian Government for the Bhutan expedition of 1865. Six bronze 3-pr smooth-bore (S.B.) guns of 2¼ cwt were turned down to 224 lb, bored

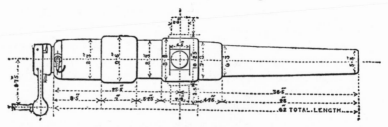

9-pr RBL gun of 1877. From S. L. Norris's *Treatise on the Construction of Ordnance.*

to 3 in, and rifled on the French system. But they were too heavy to load on a mule and a gun weighing not more than 200 lb was requested. In order to make use of the large existing stock of bronze S.B. 3-prs, a modified conversion weighing 200 lb was produced in 1866 and designated Mark II. The swell of the muzzle had been removed from the gun and it was two inches shorter in the bore. A pattern was sent to India for local manufacture and in 1867 twelve were sent to Ireland on account of the Fenian troubles. Three years later, six were dispatched to Canada to provide light equipment for the Red River Expedition. These conversions were sufficiently promising for a Mark III bronze 7-pr to be designed which was to be built new; but only two were in fact made because in 1873 it was decided that no more bronze 7-prs should be made.

Already in 1865 a steel 7-pr had been produced, which weighed 190 lb and was designated Mark I. Five equipments were made and sent to Bhutan. The Mark II, a lighter version weighing 150 lb, appeared soon afterwards and twelve were sent on the Abyssinian expedition. Its fine performance in that campaign earned it the name of the 'Abyssinian gun'. As a result of the experience gained in Abyssinia the gun was modified and became the Mark III of the same weight. A horizontal hole was bored through the cascabel for a rod to be passed through it for loading and unloading, and also to enable men to carry the gun over country impassable for animals. A Mark IV, weighing 200 lb, was produced in 1873. A much longer gun, it was primarily intended to take the place of the smooth-bore mortars in the siege train.

There were four types of 9-pr R.M.L.s: the 6-cwt Mark I S.S., the 6-cwt Mark II L.S., the 8-cwt Mark I L.S., and the 8-cwt Mark II S.S. The 6-cwt Mark I was the shortest and was meant for boat service; some being supplied to the Indian Naval Service. The 6-cwt Mark II was the equipment for Horse Artillery and the light field batteries. The 8-cwt Mk. I was for the heavy field batteries, but it proved unsuitable and guns of this type were withdrawn and modified for sea service. They were then similar to the Mark II which had been designed for this purpose.

In 1869 General Dickson's Special Committee on Shrapnel *v.* Segment Shell stated that the existing 12-pr and 9-pr common shell were of little effect against earthworks and not much good either for use in the field or for attacks on entrenchments. Their recommendation is given above. In 1870 a Special Committee, in the light of this report, recommended a rifled gun of about 12 cwt with a calibre of between 3.5 and 3.7 in. Two guns were accordingly made with respective calibres of 3.6 and 3.3 in. The former was that adopted for the heavy field batteries and became the 16-pr 12 cwt gun Mark I.

The 25-pr R.M.L. 18-cwt mentioned above, was proposed in 1871 by the Superintendent of the Royal Gun Factory for use as a light siege gun and gun of position. The decision to adopt it was taken in 1874.

Like the R.B.L.s, there were two types of 40-pr R.M.L.s with only a minor difference: the Mark I weighed 34 cwt and the Mark II 35 cwt. The difference was accounted for by a slightly different method of construction. Both pieces were for land service.

The smallest howitzer was the 6.3-in. which owed its inception to the success of the larger 8-in. and the wish to have a howitzer that would fit the 40-pr gun carriage. The 8-in. howitzer was the first piece of ordnance introduced to replace the S.B. mortars for high-angle fire, and in 1872 instructions were given that these howitzers should replace the mortars in the siege train.

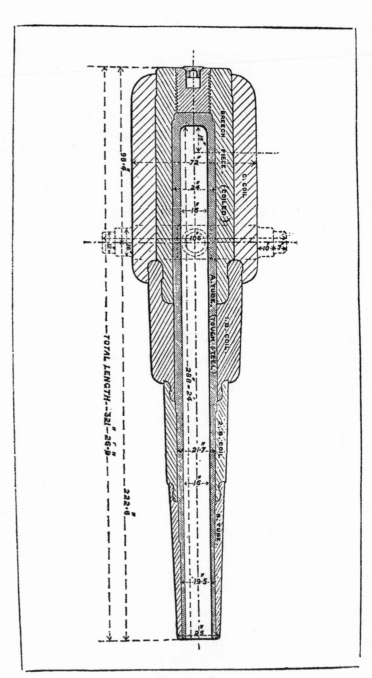

A RML gun of 1877.

The heavier guns were considerable in number and may be dealt with more summarily. They consisted of 64-prs (both new and converted from S.B.), 7-in., 8-in., 9-in., 10-in., 11-in., 12-in., 12.5-in., 13.05-in., and 16-in.; a total of ten different calibres and over thirty different marks. Nearly all were provided for both land and sea service. (It should, perhaps, be explained here that the design and supply of pieces of ordnance for both Services had been the responsibility of the Ordnance Department. When that Department came to an end in 1858, this responsibility passed to the War Office, and so remained until after Captain J. F., later Admiral of the Fleet, Earl Jellicoe became Director of Naval Ordnance in 1904. Jellicoe then managed to gain control for the Admiralty of the supply and to some extent the design of Naval guns.) The 64-pr was used by the Navy as a broadside or pivot gun and by the Army as a siege gun. A 7-in. 90-cwt gun formed part of the armament of unarmoured vessels, and a 7-in. of 6½ tons was a broadside or pivot gun for frigates. The Army had a 7-in. of 7 tons (longer than was suitable for naval purposes) as a 'battering gun' for coast defence. The Navy used 8-in. guns for vessels not sufficiently heavy to carry 9-in.; and the 9-in. was a broadside gun for the heavy ironclad ships and for the Army's defences of harbour and sea fronts. The 10-in. was proposed by Commodore Heath R.N. in 1865 as an improvement on the heavy 9-in., and H.M.S. *Hercules*, completed in 1868, received this new gun and thicker armour, and was the most powerful warship afloat. The Army then adopted it for coast defence. The great penetration achieved with the 10-in. led to the introduction of an 11-inch, with the additional object of comparing it with the 12-in. It was held on 1 October, 1870 that a comparison between the two showed the 11-in. as the better gun. It was only occasionally used by the Navy, however, and the Army added it to its coast defence armament. The biggest of the R.M.L.s was the 16-in., and four of these great guns were mounted in the battleship H.M.S. *Inflexible*, completed in 1881 and the most powerful warship ever built with a muzzle-loading armament.[35]

CHAPTER IX

The Return of Breech-Loading

IN the 1870s the great increase in the thickness of armour directed attention to the possibility of so improving ordnance that higher velocities and greater penetration could be obtained without increasing the weight and bore of the piece. An obvious source of increased power was the powder, and it was the search for this that had produced the prismatic powder. But when this powder was used in the short R.M.L.s of the period it was found that the projectile left the muzzle before all of it had been consumed and that much power was thereby wasted. Still-burning powder, indeed, scored the decks of ships.[1] The experiments carried out by Armstrong's firm with long guns and slow-burning powder (see the previous chapter) were under the supervision of Captain Andrew Noble, who was to become very eminent in this sphere. Noble (1831– 1915) served in the Royal Artillery from 1849 to 1860. In 1858 he was secretary to the Committee established to examine the merits of Armstrong's rifled guns. Two years later he retired from the Army and became a partner in Armstrong's Elswick Ordnance Company. In 1893 he was made a K.C.B. and in 1902 a baronet. He became chairman of Armstrong, Whitworth & Co. in 1900 and remained so until his death. Few men have done more for the design of ordnance and its ammunition and for the science of gunnery.

It was clear from the experiments, not only that longer guns were necessary, but that if muzzle-loading was continued these long guns could not be used on board ship, because they could not be run in sufficiently for loading. Breech-loading, therefore, had to be reconsidered. With memories of the previous breech-loaders, many were very

12. Men of the RFA in action with an 18-pr, Meteran, 1918.

13. Crews of a French 75-mm gun battery training near Bougainville, 1916.

14. Men of the RGA firing a 60–pr, near Dainville, 1916.

15. A Mark II 60–pr in action near La Boisselle, 1918. Note the length of recoil.

reluctant to contemplate a change. But the arguments from Noble's experiments were conclusive, and the muzzle-loading enthusiasts had to yield. In the words of Lloyd and Hadcock[2], 'In England there commenced probably the most extraordinary revolution that ever took place in connection with warlike material.'

To make a successful breech-loader, the first considera-tion was obviously the design of the breech action. Neither the original Armstrong arrangement nor the Krupp wedge system had been satisfactory. For the new breech-loaders the 'interrupted screw' used in France was chosen. In this system the opening of the breech screw required four motions: first, to unlock it; second, to turn it a fraction anti-clockwise to clear the threads from engagement; third, to withdraw it; and fourth, to swing it round to the side of the piece. To close the breech the above actions were reversed.[3]

In 1881 a 12-pr R.B.L. gun was produced by the Elswick firm and its great superiority over any previous gun was convincingly demonstrated. As a result it was adopted as the equipment for both horse and field artillery. To be fit for the former role, it needed to be especially light, and the weight of the whole equipment, including its limber and 34 rounds of ammunition, was accordingly limited to 33¼ cwt. Various subsequent alterations and additions, however, increased this to 38 cwt. The breech was closed by the interrupted screw, and obturation (i.e., sealing) was effected by a steel cup which was fitted to the breech screw. The piece was made entirely of steel and consisted of a tube with a jacket shrunk over it, the latter being lengthened at the breech end to receive the breech screw.[4] In practice this equipment proved too heavy for horse artillery and a new and lighter 12-pr R.B.L. gun with a simple carriage was introduced for the R.H.A. in 1894.[5]

There was nothing particularly new about the steel Mark I carriage with 5-ft wheels on which the piece was originally mounted. But in 1881 a recoil carriage for use in the field was designed at Woolwich. After extensive trials and some modification it was adopted early in 1890 for Service use as Mark II. It had a top carriage with a recoil cylinder connected to it by trunnions, and the piece was fitted into a cradle which slid in guides on the upper part of the top carriage; to the breech end of the piece was attached the piston rod of the recoil cylinder. When the piece recoiled it compressed the springs in the cylinder, allowing a backward movement of about four inches. This was not sufficient to absorb the full recoil, but it lessened the shock of firing on the main carriage, and to the latter was attached a self-acting tyre brake which checked the remainder of the recoil. The development of what was at first called the 'Elastic Field Gun' is further considered in a later chapter.[6]

There was also a 12-pr gun for boat service and this was so constructed that it could be used as a naval gun ashore. It was fitted for haulage by manpower, so that the gun's crew could pull it into any required position.[7]

A revolution in the type of ordnance used in the field owed its origin, apparently, to the very effective head cover introduced by the Turks during their gallant defence against the Russians in the war of 1877. The Russian guns were quite ineffective against it and, as a result, similar protection for infantry manning defensive works came much into favour in many armies. The counter to it seemed to lie in arranging a steep angle of descent for the projectile, which inplied a return to the howitzer.[8]

In the days of S.B. ordnance, with its limited range, all field batteries had both guns and howitzers, but the latter had dropped out of use with the introduction of rifled pieces.[9] Now there appeared to be a need for a rifled howitzer. To meet this need, Elswick produced a 12-cm. B.L. howitzer mounted on a field carriage of the ordinary travelling type and with a steel recoil cylinder. In addition to this carriage there was a wooden platform with pivot and plate for regular siege operations. The first field howitzer batteries were organized in 1896 and equipped with 5-in. howitzers, mounted on the above carriage, while siege trains were provided with a heavier 6-in. howitzer on a static mounting.[10]

National Geographic has packaged America

...to guide you through our great land.

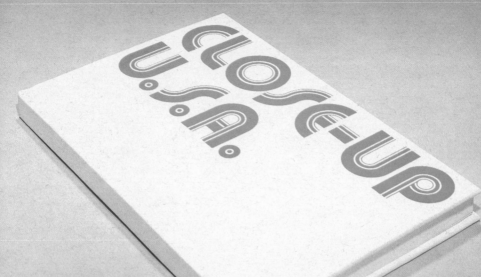

A $64⁰⁰ value in maps alone.
Yours for just $19⁹⁵!

This unique National Geographic travel planner includes:

16 full-color plastic maps, nearly two feet by three feet, regularly priced at $4.00 each . . .

- A map of the entire U.S.A.
- 15 Close-up maps, fully indexed, completely up to date, and printed on both sides — showing our nation's regions in the *largest scale* ever presented by National Geographic.
- Descriptive text and map notes covering points of interest and outstanding seasonal events in all 50 states.

A "mile-o-meter" in clear plastic that can be placed over any section of the U.S.A. map and show you approximately how far you can travel in up to nine hours of driving . . .

- at speeds of 35, 45, and 55 m.p.h.
- in both statute miles and kilometers.

An index-guide to all 15 of the "Close-up: U.S.A." maps . . .

- 200 pages, hardbound cover.
- Chart showing average temperature ranges for representative cities.

- A table of mileage distances between major U. S. cities.
- Nearly 45,000 entries of places your entire family will enjoy visiting.
- Names and locations of cities, rivers, lakes, mountains — a myriad of geographic features throughout the nation.
- State and national parks, forests, seashores, monuments, wildlife and nature preserves, recreational sites.
- Ski, scuba-diving, and canoe areas, hiking trails, caves, mines, ghost towns, Indian reservations and archeological sites, museums, historic shrines, and much more.

A handsome and practical filing case . . .

- handy size, 3⅜" x 6¾" x 10".
- made of sturdy high-grade plastic to withstand years of frequent use.
- imprinted on the inside cover with a color-coded U.S.A. map for quick and easy referral to any of the 15 "Close-up: U.S.A." regional maps.

Take advantage of this unique offer today!

The National Geographic "Close-up: U.S.A." travel planner will give you at a glance material you could duplicate only by consulting dozens of guidebooks, brochures, atlases, and directories. Nearly 70,000 man-hours, involving the talents of cartographers, illustrators, writers, and exacting researchers, have gone into the making of this package.

You'll find it an indispensable guide . . . on the road . . . in armchair travels . . . to places in the news . . . to U. S. geography in school assignments.

Because the 16 maps, "mile-o-meter," and map index-guide will be shipped *as a single package,* the price is just $19.95. This is a limited-time offer. After May 31, 1978, the price will be $23.95.

NATIONAL GEOGRAPHIC SOCIETY

Actual-size detail from "Close-up: U.S.A., Western New England"   B-1

Another innovation of the period, and a very important one, was the 'Quick-Firing' gun. It originated in the machine gun. The first effective machine gun was the French *mitrailleuse* of the 1870 Franco-Prussian War, which consisted of a number of fixed barrels and had a firing rate of 30–50 rounds per minute. The *mitrailleuse* was inevitably followed by other machine guns. Dr Gatling invented a gun which was adopted by the Royal Navy and used in a number of minor campaigns. The Gatling gun fired at the astounding rate of 600 rounds a minute. It consisted of ten barrels fixed in frames which revolved about an axis, each barrel firing in turn as it reached the firing point. The advantage of this rapidity of fire was not, unfortunately, entirely reflected in the results. A naval Gatling crew, of eighteen ratings, competed at Gibraltar against eighteen picked other ranks of The Rifle Brigade armed with Martini-Henry rifles. Over a given period the Rifle Brigade team achieved the higher score on the targets![11]

The Gatling was superseded by the more reliable Gardner. The barrels of this gun were fixed side by side and were fired by turning a crank handle. The number of barrels could be one, two, or five, and the rate of firing was 120 rounds per barrel per minute.

A still better gun was that invented by Nordenfeldt, but it was not adopted immediately by the British Services because the Gardner had only just been issued when the Nordenfeldt appeared. Eventually, however, the Nordenfeldt became not only the standard British machine gun, but it was acquired by nearly every other European nation. It had five barrels, arrayed side by side, and it was fired by moving a lever backwards and forwards. The firing rate was about 600 rounds per minute.

The Nordenfeldt was followed by the marvellous Maxim automatic machine gun, which, with its Vickers derivation, became the British heavy machine gun until recent years. This gun had only one barrel and, except for keeping the trigger pressed, no action was required of the firer.

All these weapons were of rifle calibre. In 1877 Nordenfeldt produced a 1-in. gun, worked in the same fashion as his rifle-calibre machine gun, which was intended for defence against torpedo boat attack, and fired solid projectiles at the rate of about 216 a minute. This gun was adopted by the Royal Navy as an anti-torpedo boat armament.

A rival to this Nordenfeldt gun was the Hotchkiss 'revolving cannon', which was chosen by the French and some other foreign navies. The Hotchkiss had not such a rapid rate of fire as the Nordenfeldt, but it fired shells instead of solid projectiles. Of several calibres available, the favourite was 37 mm. (1.46 in.) with a shell of just over one pound.

The increasing size of torpedo boats led to a need for a heavier weapon than either of the above. With both the Nordenfeldt 1 in. and the Hotchkiss 37 mm., metal cartridge cases had been used and these paved the way to their use with heavier calibres. The metal cartridge case was the key to rapid fire because it enabled a fresh charge to be loaded safely immediately after firing without sponging or washing out. In addition, it avoided the delay of removing the old and inserting a fresh primer because metal cartridges carried their own means of ignition.[12] Thus the term 'Quick-Firing' (Q.F.) came to be applied to an equipment having a metal cartridge case to contain the charge and seal the gas, whether the ammunition was fixed (i.e., projectile fixed to the cartridge case) or separate; whereas 'Breech-Loading' (B.L.) implied an equipment which used bags to contain the charge and in which sealing was effected by fixing a pad to the breech screw.[13]

Both Nordenfeldt and Hotchkiss designed and built 3-pr and 6-pr Q.F. guns to counter the big torpedo boats, with a respective rate of firing of 30 and 25 shots per minute. The Admiralty adopted both the Nordenfeldt and the Hotchkiss 6-prs, but of the 3-prs they decided that the Hotchkiss design was the better of the two. All these new guns were fired by a trigger and the empty cartridge case was removed by an extractor.

But to get effective use from a Q.F. gun it was necessary to have a suitable mounting. On board ship the mountings were necessarily fixed and there did not appear to be any great problem. The 6-pr Q.F. guns were given what was known as a 'non-recoil' mounting. There was, however, a considerable strain on the structure of the ship and for the 3-pr Q.F. gun a recoil mounting was designed and constructed by Sir W. G. Armstrong, Mitchell, & Co. This was so successful that a mounting of a similar type was produced for the 6-prs.[14]

Q.F. guns were now recognized as so essential for ship armament that a much bigger piece, the 40-pr, was submitted in 1886 by the Elswick firm for Admiralty trial. The calibre was increased to make it a 45-pr and it attained fame under its eventual designation of '4.7 quick firing gun'. The first order for the 4.7-in. Q.F. guns was given on 16 November, 1887, and by 1890 it was a staple armament for all cruisers and a secondary armament of battleships.[15]

In 1890 a still heavier Q.F. gun, a 6-in., was introduced into the Royal Navy, and in the same year the new smokeless explosive, 'cordite', was taken into use by the British Services. At about this time the difference between fixed and separate ammunition in speed and ease of firing was tested on a 4.7-in. naval gun. It was found that if the ammunition was ready by the gun, it was possible to fire about one round per minute faster with fixed ammunition than with separate; but that with the latter the work of supplying ammunition was lighter and therefore faster. With heavy ammunition, the greater speed of supplying the separate type would cancel out the greater speed of firing with the fixed; separate ammunition would therefore be preferable for the larger guns.

The advantages in 1890 of a Q.F. field gun were not so apparent. Although enthusiasts were not wanting to urge the need for a light rapid-firing equipment, it was officially considered that the existing conditions for field artillery warranted no such radical departure from current practice.[16] If Q.F. guns were provided for field batteries, the metal cartridge cases would introduce a large proportion of dead weight into the permissible loads carried by gun limbers, ammunition wagons, and reserve columns. The limber of the 15-pr B.L. gun could carry 38 rounds, but for a similar Q.F. gun the limit would be 31 rounds. Rapid fire, it was considered, would be required only on rare occasions, such as a cavalry attack on the guns or the appearance of a fleeting target. In addition, it was apparent that a Q.F. field gun would show little advantage over the B.L. type unless a full recoil carriage could be provided.[17]

The smokeless cordite met one of the needs of Q.F. guns, because rapidity of fire was of little value with gunpowder unless the wind was sufficiently favourable to disperse the smoke quickly. Besides being smokeless, cordite was very safe to handle; it could, indeed, be held in the hand and lighted without fear of explosion. While cordite was replacing gunpowder as a propelling agent, high explosive was rivalling gunpowder as a burster for a shell. The French were the first to adopt H.E., using 'melinite' which is a product of picric acid and which produces a very violent detonation.[18]

Cordite had other advantages. Experience having shown that the 12-pr common shell had no effect on earthworks, there was a demand for a field gun firing a heavier projectile and for 'one shell and one fuze'. A Committee was set up in 1892 to consider the question. Cordite, a much more powerful propellant than gunpowder, had just been adopted, and there was therefore an opportunity to increase the weight of the shell without making any radical alteration in the gun. The Committee therefore recommended that the heavy 12-pr should be converted into a 15-pr. This was accepted, and at the same time the demand for 'one shell and one fuse' was met by abolishing common shell from the ammunition of field artillery, leaving only shrapnel.[19]

Little advance in field gun recoil control was made until just before the Boer War. To the Mark I 12-pr gun carriages brakes and drag shoes were fitted, but this was only a partial solution, since, although the equipment was stopped from running back, there was inevitably a violent kick which entailed relaying the gun. The Mark II, with short recoil cylinder, had not been generally adopted. In 1899, however, the axle spade was being introduced.[20]

CHAPTER X

The Boer War, 1899-1902

THE artillery equipments used in the Boer War could be divided into (*a*) Horse and Field, (*b*) Heavy and Medium, and (*c*) Mountain.

In 1899 the Horse Artillery was armed with the 12-pr B.L. gun of 1894 which fired a 12½-lb shrapnel shell with a time fuse allowing for a range of 3,800 yards. The weight of the equipment was 31 cwt. Recoil of the carriage was controlled by the spade and spring arrangement.[1] Hinged to a bracket bolted to the bottom of the carriage and bearing against the underside of the axletree, was a telescopic spring case with a spade at its lower extremity. When this was lowered into the firing position the spade was in contact with the ground slightly behind the axletree. A second spring case was fixed obliquely in the rear portion of the trail and connected by wire ropes to the spade. When the gun was fired the carriage recoiled and forced the spade into the ground. As the carriage moved back the wire ropes were pulled forward, drawing out a tension rod and thus compressing the springs in the trail case. When the energy of the recoil had been absorbed these springs returned the carriage to the firing position. The function of the springs under the axle was to cushion the shock to the equipment when the gun was firing on hard ground.[2]

The Field Artillery had the 15-pr 38-cwt equipment which had been converted from the heavy 12-pr. It fired a 14-lb shrapnel shell with a time fuse up to 4,100 yards. (In 1900 a new fuse was sent out to South Africa from England which increased the range of both this and the 12-pr by 1,800 yards.)[3] The gun carriage was fitted with the same spade and spring attachment as described above. Some of the equipments had the Mark II carriage mentioned in the last chapter with short recoil cylinder. The dragshoe brakes of these were replaced by spade and spring attachments. The recoil cylinder was a hydraulic buffer and the top carriage in which the gun recoiled consisted of two steel guides, connected together in front and rear and supported at the back by the elevating gear. On firing, the gun, supported by its trunnions and a rear bracket, slid backwards along the top carriage, drawing a piston rod out of the recoil cylinder, the oil in which was forced by pressure from the rear to the front of the piston. At the same time springs, which were housed in the same cylinder, were compressed. When the piston had completed its stroke the springs returned the gun to the firing position.[4]

In addition to the 15-pr gun, the Field Artillery was also equipped with the 5-in. B.L. howitzer which had a range of 4,800 yards and fired a 50-lb lyddite shell. (Lyddite had been introduced in 1898 in replacement of gunpowder as the bursting charge of common shell for all B.L. pieces of over 4.7-in. calibre.) This had a carriage somewhat similar to that described above, but it had not been modified and still retained the dragshoe brakes instead of being fitted with spade and spring cases. The arrangement was different in that the springs, instead of being inside the buffer cylinder, were mounted in separate boxes on either side of the hydraulic buffer.[5]

The Medium and Heavy Artillery in the South African War consisted of the 4.75-in. Q.F. gun, the 5-in. B.L. gun, the 6-in. B.L. howitzer, and the 9.45-in. B.L. howitzer. The 4.75-in. Q.F. gun was so called because it used fixed ammunition, but it did not have a full recoil carriage. It fired a 45-lb shell, both H.E. and shrapnel, and had a percussion range of 10,000 yards. The carriage, which was improvised, had a hydraulic buffer giving a short recoil of 12 inches, two spring boxes, and the spade and spring fitting.[6]

The 5-in. B.L. gun fired a 50-lb H.E. shell to a percussion range of 8,700 yards. The improvised carriage for this gun had been converted from that used for the 40-pr R.M.L., specifically to mount it for service in South Africa. It had nothing better than dragshoe brakes to check recoil.

On the limber of the 12-pr was an ammunition box divided horizontally into two compartments; of which the upper one was subdivided to carry fuses, tubes, and a tray of small stores, while the lower contained nine four-round ammunition carriers and a second tray of small stores. The 15-pr limber was similar except that the box took eleven ammunition carriers. The ammunition box on the limber of the 5-in. howitzer was constructed in a different fashion. It was divided into partitions to take sixteen projectiles and sixteen cartridges (the latter being carried in tin boxes) together with fuses, tubes and small stores. The projectiles were placed on their sides with bases towards the front and resting against wooden blocks. Leather straps were attached to them for ease of withdrawal.[7]

The Boer War brought about a revolution in artillery tactics. The practice had been to site guns in the open because they were not equipped to shoot and hit from under cover. But when employed against the well-concealed Boers, this practice led to very heavy casualties. The current doctrine laid it down that the artillery should first silence the enemy artillery and then neutralize that portion of the hostile position against which the infantry were to attack. It was frequently impossible to see either of these targets, and the guns came under heavy shell and rifle fire from the concealed enemy.[8]

The new smokeless explosives allowed guns to be concealed; but it was no longer possible for them, if concealed, to fire over open sights. A new system of sighting was therefore urgently needed, and an arrangement was improvised called the 'Gunner's Arc'. The ordinary foresight was replaced by a strip of wood fastened at right angles to the piece, with holes bored in it at half-degree intervals. The centre hole was in the same position as the old foresight, and into this a matchstick was inserted as the new foresight. An aiming point was selected on the crest of the hill in front which was approximately in line with the target and the sights aligned on it. On top of the crest was an observer with a view of the enemy's position. The gun having fired, the observer would give a correction of, say, 'left three degrees'. The matchstick would then be moved six holes to the right and the sights re-aligned on the aiming point. Other more sophisticated systems soon followed, but the 'Gunner's Arc' was remarkably effective.

The British field artillery equipment was clearly not up to modern requirements; particularly as it was reliably reported that the French field artillery had been rearmed with a Q.F. gun capable of firing 20–30 rounds per minute. (This, in fact, was the famous 75-mm.) While the Boer War was still being waged, the Director General of Ordnance, General Sir Henry Brackenbury, G.C.B., P.C. (1837–1914), an eminent writer on military subjects, was directed to advise on the re-equipment of the Royal Artillery. Brackenbury's drastic recommendation that the whole of the existing equipment should be replaced within three years was accepted.

But although Brackenbury had a clear idea of the type of equipment required, no British firm could design and manufacture a sufficient quantity within the time required. Brackenbury, however, found what he wanted in a remarkable field gun made by a German firm in Düsseldorf, and designed by an engineer named Ehrhardt. It was a 15-pr with a long-recoil top carriage which could equal in rate of fire that attributed to the new French Q.F. field gun. Brackenbury immediately ordered sufficient of these equipments for eighteen batteries, together with 500 rounds per gun.[9] The shell weighed 14 lb and the time fuse allowed a range of 6,400 yards.

The Erhardt gun had the wedge breech action in which a slightly tapered wedge slid in a slot across the breech. The wedge had a semicircle cut from its extremity so that it corresponded with the bore when the breech was open.[10]

This was the first equipment in the Royal Artillery with a full-recoil carriage. It was not only a Q.F. gun, but it eliminated the labour shouldered by generations of gunners.

Pushing the guns back into their firing positions had been an exhausting business, and, as related by Michael Glover in *Wellington as Military Commander*, Colonel Frazer of the Royal Horse Artillery wrote that the French were able to re-form under cover of the ridge in their front because they were 'covered, in great measure from the fire of our guns, which by recoiling, had retired so as to lose their original and just position'. In the stiff soil, the fatigue of the horse artillerymen was great, and their best exertions were unable to move the guns again to the crest without horses; though to employ horses was to ensure the loss of the animals.[11]

The main feature of the equipment was, of course, the cradle with hydraulic buffer and spring recuperator. The cradle was in the form of a trough with longitudinal ribs at the top on either side, on which slid the guides of the gun. When the gun was fired it recoiled, drawing the buffer cylinder off the piston rod, which was attached to the front of the cradle. The glycerine, with which the buffer was filled, passed from front to rear of the piston through grooves in the cylinder. The recuperator, consisting of a single bank of springs, encircled the buffer cylinder. As the gun recoiled it compressed the springs, which, when the energy of recoil had been entirely absorbed by the buffer, returned the gun to the firing position.

The new travelling carriage was of much more complex design than anything that had gone before. Wheels, axletree, and trail, the historic components, were of course included, but the new carriage constituted, indeed, the greatest revolution in this essential part of artillery equipment since the introduction of trunnions in the fifteenth century.

The construction of the Ehrhardt gun was different from any used for British ordnance before or since. The tube and its jacket were each formed from a solid ingot into which a mandril was forced when red-hot. The result of the process was to leave the inner layers of the metal in a state of initial compression so that stresses were distributed evenly throughout the piece.[12]

When the re-equipment of the Royal Artillery with

British-designed ordnance took place, these Ehrhardt batteries were handed over to the Territorial Army and to some units overseas.[13]

The Mountain Artillery weapon was still the 2.5-in. R.M.L. 'screw' gun, firing a 7-lb shell. With common shell its range was 4,000 yards and with shrapnel 3,300 yards.[14] These guns used black powder, and their appearance in South Africa, where gunners had become only too conscious of the necessity of concealment, caused some consternation.[15] The failure of this gun in South Africa led to the rearmament of mountain batteries in India. Trials of a 10-pr jointed B.L. gun, already under test, were accelerated. It was clearly superior to the 2.5 in. gun but its carriage was no more advanced. Recoil was still controlled, as in the former equipment, by a 'check-rope' round the trail. Since the full-recoil carriage had already been introduced with the Ehrhardt gun, this equipment was decidedly behind the times. It was, however, the best available, and by the end of 1903 all mountain batteries had been re-equipped with it. The two parts of the gun could be assembled without noise. The carriage, which could be easily taken apart to form loads for pack transport, consisted of trail, axletree, wheels, elevating gear, sighting gear, and check rope. The trail was made of two parallel steel plate brackets, the axletree was a forging of solid steel and the wheels had single spokes and no 'dish'. The gun weighed 404 lb, had a calibre of 2.75 in., and was 6 ft 4½ in. long. Initially, the only ammunition used, apart from a certain amount of starshell, was shrapnel, but common shell was included later. The charge was of course cordite (after the South African fiasco!) and the maximum range was 6,000 yards (though ranges of up to 4,200 yards only were engraved on the sights). As compared with its predecessor, there was a 50 per cent increase in range, and with breech-loading and no black smoke to obscure visibility, the 10-pr could be fired much more rapidly than the 2.5 in. As with the field guns in South Africa, it became the practice to fire from behind cover. But since the necessary instruments for this had not yet been provided, the excellent improvised system of the 'Gunner's Arc' or 'gun arc' was adopted.

The 10-pr equipment had not been long in use before the inevitable demand arose for a long-recoil carriage. Trials were carried out and in due course a new carriage was provided with hydraulic buffer and springs; the resulting equipment (which required six pack mules as compared with five for the 10-pr) was renamed 2.75-in. B.L., but batteries were only receiving it during the First World War.[16] The trail of this new carriage was in two parts, the rear portion in the form of a box tapering from front to rear with a spade at the back, and the front portion box-shaped in rear but having side brackets in front. Special fittings connected the two parts. Tubular shafts, together with fittings for them on the trail, were provided for mule draught when required, and a bullet-proof steel shield gave protection to the gunners. The cradle was of steel U section, and the hydraulic buffer and springs were carried in it and below the gun.

The arrival of the 10-pr gun on the North-West Frontier was a nasty shock for the Pathans. Previously the black smoke from the discharge of the 2.5-in. had given them time to take cover before the shell arrived. Now there was nothing to warn them before the shell burst.[17] British Mountain Batteries were issued with the 10-pr before Indian Batteries, and the first occasion of its use on active service was in Aden in March 1903. The 6th British Mountain Battery destroyed the towers of the village of Chali which was harbouring some of the trouble-makers.[18]

Shafts had been used for Field Artillery gun carriages up to 1895, but in that year they were replaced by pole draught. To design the ideal gun carriage was no easy matter. It had to provide adequate support for the piece both on the move and in action. For movement it had to be light enough to be pulled and reversed easily, and to pass over obstacles; yet it had to be strong enough to carry its load and to support the piece when it was fired. The load, on the other hand, had to be as light as possible and properly distributed, the wheels of as large a diameter as convenient, and the axletree of as small a diameter as would give adequate strength. The wheel diameter, previously 5 ft, was reduced to 4 ft 8 in. Springs were not generally adopted since they introduced some weakness and entailed making the carriage higher.[19]

In order to suit military vehicles for cross-country travel, their wheels were constructed in a particular fashion. To enable them to withstand the side blows inflicted by the axletree on the nave, 'dish' was given to the spokes; that is to say, they were not perpendicular to the nave but were inclined outwards from the carriage. But if a wheel with such spokes was fitted vertically, the working (or supporting) spoke would be inclined to the vertical, and would prove a source of weakness. 'Hollow' was therefore given to the axletree arm by inclining its axis downwards so as to make the working spoke nearly vertical. At the same time the underside of the axletree arm was kept as nearly horizontal as possible and the inclination of the axis was obtained by making the arm cone-shaped. The working spoke was, however, not quite vertical on flat ground, thus ensuring that it was truly vertical on the average slant towards the downhill wheel, when the pressure on the spoke was greatest. The inclination of the spoke away from the vertical was called the 'strut'. The sole of a hollowed wheel was bevelled so as to bring the whole breadth of the tyre in contact with the ground. The bevel would tend to make the wheel roll outwards from the axlebox and the carriage consequently hard to pull. To counteract this, a slight lead was given to the axletree arm, so that the wheels were 'toed in'. Hollow and lead made up the 'set' of the axletree arm.[20]

CHAPTER XI

The Q.F. Era

IN 1901 British gun-making firms were invited to submit designs to meet the conditions laid down by a committee in the light of experience with the Ehrhardt gun. The following year, specimens submitted by these firms underwent trials; none was entirely satisfactory. The committee therefore asked the firms concerned if they would take the best features of each, and jointly produce a gun incorporating the Armstrong wire-winding, the Vickers recoil arrangement, and the Ordnance factories' sighting and elevating gear and method of carrying ammunition.[1] This was agreed, and in 1903 four batteries equipped with two types of Q.F. gun, a 13-pr and a 18-pr, were ready for trial. Both these equipments met Service requirements, with the result that the 13-pr was adopted for the R.H.A. and the 18-pr for the R.F.A. By 1906 the regular divisions of the field force had been equipped with the new guns.[2]

The wire-wound system of gun construction had been invented by James Longridge. In the old method, which it superseded, pieces had been built of concentric tubes, the outer ones shrunk onto the inner. In Longridge's system a steel ribbon, $\frac{1}{16}$ in. thick and $\frac{1}{4}$ in. wide, was wound round a single tube in successive layers, with gradually decreasing tension. It was found that a wire-wound piece could be made lighter than a built-up one of the same strength, and the manufacture was thus considerably cheaper.[3]

The famous 18-pr Q.F. gun of 1903 was a 3.3-in. calibre piece of wire-wound nickel steel with two guide ribs, one on either side, for nearly the whole of its length. It recoiled, sliding on the guide ribs, in a bronze ring cradle, checked by a hydraulic buffer placed above the gun and surrounded by telescopic springs. The breech action was the interrupted screw with single motion. On pulling the breech lever, the breech screw was turned through a quarter circle, withdrawn, and swung round to the right; at the same time the empty cartridge case was ejected. The 13-pr gun was of identical design. Both guns could be fired at a rate of twenty aimed rounds a minute and were very accurate.[4] With such a revolutionary and rapidly produced equipment, there were certain inevitable defects. The carriage, for instance, proved insufficiently steady in firing, so that the gun had to be relaid after each round.[5] The advisability of using H.E. with these guns was debated, but because a field gun could be knocked out by a direct hit on its shield with shrapnel, it was decided that there was no necessity to accept the complication of having two types of ammunition[6] – a decision to be regretted later. A number of the 15-pr B.L. guns were given long-recoil carriages and were known as 15-pr B.L.C. (Breech-Loading Converted). They were issued to field batteries of the Territorial Force.[7]

The pioneer of all Q.F. equipments was the excellent French 75-mm. of 1899. To obtain the high velocity required to give a flat trajectory, the piece had the unusual length of 36 calibres. The object of the flat trajectory was to get an extensive shrapnel zone; but the necessary velocity resulted in a high muzzle energy, and to keep the gun steady against this the diameter of the gun carriage wheels was reduced to four feet. The piece recoiled on the carriage for 48 inches. There was the usual hydraulic buffer to check the recoil but, unlike the British field equipment, the gun was returned to the firing position by compressed air acting on the buffer piston; the air chamber being a prolongation of the buffer. In addition to the recoil on the carriage, the carriage itself was secured by a narrow pointed spade and by dragshoes formed by dropping tyre brakes under the wheels. To bring the gun into action or to shift it onto a fresh target, the trail had to be raised to shoulder height to allow the brake blocks to drop under the wheels. It was then lowered to bring the gun into the line of fire. This operation, which was called *abatage*, had to be carried out whenever the gun was moved more than three degrees. The breech action was the eccentric screw. This was a screw at the breech, mounted eccentrically, which had a hole at one side of its centre of the same diameter as the bore of the gun. A half-turn of the screw sufficed to make this hole coincide with the bore, and another half-turn closed the breech by bringing the solid part of the screw opposite the bore. Originally the equipment had no shield, but one was added in 1902, and the axletree seats were removed to compensate for the extra weight. (Three gunners were then carried on the gun limber and three on the wagon limber.) The French field battery consisted of four guns and twelve wagons. Of the latter, half were unlimbered with the battery in action, the others were sited about a mile in the rear. The gun was supplied with both shrapnel and H.E.[8]

The German 1905 Q.F. equipment was a conversion from the 15-pr of 1896, the recoil of which had been controlled by a spring spade. In the new equipment the gun was mounted on a long-recoil carriage with a shield, and was designated M96n/A. The breech had the Spandau single-motion wedge action, in which a wedge slid in a slot across the breech to open or close it. A hydraulic buffer allowed a recoil of 44 ins and the gun was returned to the firing position by 'running up springs'. In 1910 this gun's ammunition consisted of 80 per cent shrapnel and 20 per cent H.E.; but the Germans were experimenting with a universal shell, the Ehrhardt combined shell which had a shrapnel body with a H.E. head. When this burst in the air, the bullets were blown out in the normal fashion, while the head went on and exploded on impact (thus, incidentally, providing a useful means of observing and correcting). This shell, somewhat modified, was ultimately adopted as a 14.3-lb projectile for the German 4.2-in. Q.F. howitzer. The 15-pr was also the equipment of the German Horse Artillery, but lightened by removing the axletree seats. In the Field Artillery five gunners were carried on the gun carriage and limber, and two on the wagon. A horse or field battery consisted of six guns and wagons; three batteries constituted an artillery brigade and there were two artillery brigades in an artillery regiment.

The Austrian 1905 pattern Q.F. field gun was very similar to the 15-pr Ehrhardt gun supplied to the British Army; but the Austrian gun was unique in that it was made of hard-drawn bronze. It was probably only in the Austrian Army that bronze, the old 'brass', lasted into the Q.F. era.

The Russian field gun, used in the Russo-Japanese War, was a rather unusual equipment. It had a high muzzle velocity and its shooting was considered satisfactory during the war, but it had a number of disadvantages. The recoil arrangement incorporated a hydraulic buffer and springs, but they were mounted in the trail and allowed a recoil of only 36 inches. The carriage had a spade and tyre brakes as well, but it was very unsteady and the springs, consisting of rubber discs separated by washers, gave considerable trouble. The sights were on the gun, instead of being on the carriage, and recoiled with the gun. There was no shield to give protection against small arms fire and shrapnel, and the equipment was far too heavy for the country in which it was employed, seriously reducing its mobility. The limitations of the gun were known before the war started and a new pattern had been designed in 1902 and modified in the following year.

The 1901-pattern field gun used by the Japanese was far worse than the Russian piece. Its shooting was poor, it had an obsolete carriage with no recoil control, other than spring spades, and it had no shield. After the war it was rapidly replaced by a 14.3-pr Q.F. equipment with shield and long-recoil carriage.

Artillery equipments of many nations were met by the standard designs of the big armament firms. The two major suppliers for foreign customers were F. Krupp of Essen in Germany and Schneider, Canet et Cie of Le Creusot in France. Krupp made four types of field equipment in 1910, which could be modified in accordance with customers' demands, and which weighed respectively 18 cwt, 19.5 cwt, 21 cwt, and 22 cwt; the last being a high velocity gun. The basic calibre was 75-mm. The Swiss field gun of 1903 was a modification of the second type with a narrow track to fit it for Swiss mountain roads. Italy had the second type; Romania chose the heavier third type. Denmark had a gun similar to that of the Romanians; so also did Sweden, though somewhat lighter. Holland had the lightest equipment, but rather heavier than the standard. Argentina and Brazil bought the light equipment, but Brazil had it modified to make it still lighter. Belgium's gun was much the same as that of Sweden, whereas Turkey selected a pattern very similar to that of the Swiss.

Unlike Krupp, Schneider in 1910 was manufacturing only one pattern of field gun, a 75-mm., 31.4 calibres long, and weighing 20 cwt in action. The breech action was very similar to that of the British 18-pr. Schneider field guns equipped the artillery of, amongst others, Spain, Portugal, Greece, Bulgaria, Serbia, and Persia. Like the Krupp guns, they could be modified to meet customers' requirements. The Greeks held a competition in Athens in 1907 in which Armstrong (the Elswick Ordnance Co.), Ehrhardt, Krupp and Schneider took part. Schneider won on account of their gun's superior steadiness and smoothness.

The American 1905-pattern field gun was a modified version of the Ehrhardt purchased for the British Army, but the breech action was a single-motion interrupted screw. In 1910 it had not yet reached all batteries, several being armed with an earlier 15-pr Ehrhardt. Ammunition was the usual shrapnel and H.E., and with the former the 1905 equipment had a range of 6,500 yards.

Howitzers presented an entirely different problem to that of guns. In 1910, rather surprisingly, no country had as yet produced a Q.F. howitzer of its own official design. The minor powers were buying standard models, generally from Krupp and Schneider.[9] Design of a suitable piece was not, indeed, easy. When the long-recoil principle was applied to a field howitzer, the considerable angle of the maximum elevation limited the distance to which the piece could recoil. Recoil length had to be shortened, therefore, by increasing the resistance to the buffer piston as the elevation increased.[10]

There were two main systems of constructing howitzers, one designed by Krupp and the other by Ehrhardt, the latter adopted by Schneider. The Ehrhardt system had a controlled recoil mechanism which automatically increased the buffer resistance to shorten the recoil as the howitzer was elevated. (A shortened recoil could be accepted because the higher the elevation the more the piece was recoiling towards the ground, rather than to the rear, and the less the horizontal pressure on the carriage.) Krupp, on the other hand, put the cradle trunnions under the breech or even further to the rear so that the elevation of the howitzer did not affect the clearance and there was room for the firm's standard 43-in. recoil. The Krupp weapon was, indeed, a gun/howitzer and was used more often as the former, unless high-angle fire was particularly needed. Both systems had, however, certain weaknesses, with the result that by 1914 Schneider and Krupp had each adopted a combination of controlled recoil and rear trunnions.

Switzerland adopted the 4.7-in. Krupp howitzer, whereas Spain opted for the 4.7-in. Schneider. Bulgaria equipped nine four-gun field batteries with 4.2-in. Schneider howitzers. Japan purchased the 4.7-in. Krupp but in a rather lighter version. Krupp supplied a 4.7-in. howitzer to Turkey, but the Turks specified controlled recoil (which at full elevation was 23 in.).

The first country to produce its own howitzer was France. The equipment was the Rimailho heavy Q.F. field howitzer of 6.1-in. calibre (155 mm.). Because of its weight the howitzer travelled on a special two-wheeled cart, its cradle remaining on the gun carriage. It had rear trunnions, allowing a recoil of 60 ins, and it was anchored by dragshoes under the wheels. A battery consisted of two howitzers, four wagons, and eight first-line wagons. The cartridge was divided into seven charges and the standard projectile was 95-lb H.E., though a few shrapnel were also carried. The ordinary full-charge range was 5,400 yards, but there was an extra charge carried for $\frac{1}{3}$ of the shell, which increased the range to 6,600 yards. In addition to this heavy field equipment, the French had a

4.7-in. (120-mm.) Schneider howitzer, and this type was also bought by Russia, Serbia and Greece.

The German Krupp-built Q.F. field howitzer was a very neat-looking equipment of 4.2-in. calibre (105-mm.) firing a 31-lb 'universal' shell. It had controlled recoil combined with rear trunnions and weighed 22½ cwt. The Belgians produced a howitzer with semi-rear trunnions and constant long recoil. Sweden adopted a new Bofors howitzer with controlled recoil. Switzerland chose a 4.7-in. Krupp howitzer with a combination of rear trunnions and controlled recoil. Italy chose a heavy 5.87-in. (147-mm.) with a motor limber for road traction. Chile bought twenty-four Krupp howitzers similar to the official German pattern; Romania bought sixty of these and also eight Schneider 6-in. howitzers.

The British Army was equipped with its new 4.5-in. Q.F. field howitzer shortly before the outbreak of war in 1914. It was a controlled recoil equipment with a wedge breech action and it proved to be an excellent and very accurate piece. Its ammunition consisted at first of 70 per cent shrapnel and 30 per cent H.E.; but the issue of shrapnel was stopped very early in the First World War.[11]

There was considerable difference between the various methods of mounting piece and recoil gear in Q.F. equipments. In general the buffer was so arranged as to be parallel to the piece, whatever the angle between the latter and the main carriage. This was done by mounting piece and recoil gear in a cradle which had slides for the piece and trunnions to connect it to the main carriage. The only exception to this practice was the above-mentioned Russian 1900-pattern field gun in which the buffer was parallel to the trail instead of to the piece. Generally the buffer was mounted below the piece, as being the simplest arrangement; but in the British 18-pr and 13-pr guns the buffer was above the piece and the short ring cradle was supported by trunnions on an intermediate carriage which could traverse horizontally on the main carriage. The object of this arrangement was to get the gun as low as possible.[12]

Krupp's cradle, with the piece on top, had been modified

from the Ehrhardt pattern fitted to the British equipment made by that firm. It was a steel tube containing the buffer and springs with slides on the top for the steel blocks formed on the outer tube of the gun. The cradle had a single vertical trunnion pivoted on the axletree, thereby avoiding the use of an intermediate carriage with its horizontal trunnions.[13] German, Austrian, and Ehrhardt cradles were long and, respectively, box-shaped, cylindrical, and U-shaped.[14]

For Q.F. mountain guns, all continental nations in 1910 depended on equipments produced by private manufacturers. Spain, Portugal, Serbia, Bulgaria, and Greece had, as might be expected, purchased Schneider guns. Equally expectedly, Italy, Switzerland, and Turkey bought their mountain guns from Krupps. The Schneider mountain guns were reduced copies of their field guns, but General Danglis of the Greek Army produced a modification which resulted in Greek mountain batteries being equipped with the far more powerful Schneider-Danglis gun, which fired a 14.3-lb shell – heavier than that used by either the Dutch or Swiss field guns. This Greek gun was in two pieces, and the whole equipment constituted five mule loads.

Except for a divided trail and certain modifications for quick assembly, the Krupp mountain gun was similar to the firm's field gun. Krupp manufactured also a 4-in. Q.F. mountain howitzer firing a 25-lb shell, which was in two pieces, each weighing 265 lb. The recoil gear was the same as that of the Krupp field gun but with rear trunnions, and there was a constant long recoil of 27 ins. In 1910 Krupp produced a more powerful 4.2-in. version.

Guns and howitzers made at this period in two pieces were not jointed, but the breech ring and mechanism were carried separately.

The Austrians had a 4-in. Q.F. mountain howitzer, but this was not a pack equipment. On the move the carriage was empty and attached to a limber drawn by two horses, with the cradle and piece following, each on a two-wheeled cart. On rough ground the wheels could be removed and the vehicles drawn as sleighs.[15]

An artillery surprise of the South African War had been the heavy guns of the Boer Army and the unpleasant effectiveness of their fire. Consequently it was decided that the British Army should have a heavy artillery brigade by converting three siege companies. The brigade was formed in 1903, each of its batteries being equipped with four 4.7-in. guns. This, however, was only a makeshift, for the 4.7-in. had a flat trajectory and an unwieldy carriage and was thus far from ideal for field use.[16]

In 1902 a committee, set up to select a new equipment, chose a promising Armstrong 60-pr B.L. gun. After a long series of tests the choice was approved.[17] In 1906 a four-gun battery, armed with the new 60-pr, was allotted to each infantry division, with ammunition consisting of 70 per cent shrapnel and 30 per cent H.E.[18]

The trail of this equipment consisted of two parallel side brackets connected by a top plate and transoms at the front, the rear, and an intermediate position. On top of the trail was a saddle which was an intermediary between the trail and the carriage body and enabled the latter to be traversed, as well as allowing a backward and forward movement. Mounted on the carriage body was the cradle, made of front and rear collars joined by two side bars. The collars had four openings, of which the largest and lowest was for the gun, the top small centre one was for the hydraulic buffer, and the two larger openings on each side of the latter accommodated the spring recuperators. The buffer cylinder was fixed to the upper portion of the cradle, with the piston rod attached to it and moving with the gun. The length of recoil was 60 inches.[19]

Two old equipments were retained for use in the field force as siege pieces. One of these was the 6-in. B.L. howitzer of 1896, of which there were four batteries, and the other was a 9.45-in. howitzer made by the Skoda Works in 1900, a rather unsatisfactory weapon with which one battery was equipped. The 9.45-in. was split into two loads for movement; the bed, with carriage body and gears, being one load, and the cradle, with howitzer and recoil gear, the other.[20]

CHAPTER XII

Naval Development up to The First World War

WHEN Captain J. F. Jellicoe became Director of Naval Ordnance in 1904 he managed, as we have seen, to wrest from the War Office some measure of control over the guns for the Navy. The Royal Navy had certainly good reason to complain of the handling of their artillery affairs, for many of the rifled guns listed in Chapter VIII had been far from suitable for the sea service. The introduction of rifled guns had, indeed, approximately coincided with the introduction of armour plate as a protection for warships; the Navy's primary need, therefore, was for a gun which would penetrate the thickest of such armour.

The first cause of complaint was the Armstrong breech-loaders introduced into the Fleet in 1860. Two sets of trials carried out in 1859 had been a dismal failure. In the first of these an Armstrong rifled breech-loader, firing shot of variously wrought iron, cast iron, and steel against the 4-in. armour-plated sides of a floating battery, at ranges decreasing from 450 to under 50 yards, failed to achieve any appreciable effect. These results, however, were not accepted by the Special Ordnance Committee and, with minor adjustments of shot and charge, the gun was accepted for the Navy without further trials as the Armstrong 40-pr.[1] In the second trials another Armstrong R.B.L., described as a 'special gun of large calibre', fired at the same battery over the same ranges with cast iron and steel shot of 78–100 lb without making any impression on the 4-in. plates. Nevertheless the War Office ordered a heavier gun of the same 7-in. calibre as the standard naval heavy gun.[2]

Something of a naval revenge came in September, 1861 when trials against armour plating between the new 7-in. and the old S.B. 68-pr (see Chapter VII) showed the latter to be superior.

Apart from their ineffectiveness against armour, it seems that at this period the art of making large pieces without flaws had not yet been acquired. Flaws, in fact, must have been very frequent, because the Gunnery Captain of H.M.S. *Cambridge* said, 'None of the Armstrongs that I have seen have been without flaws. It will be a long time before it bursts, but a gunner does not like to stand along-side a gun with a few cracks in it. I have always felt that I would rather the manufacturers of the gun should fire it than I should'.[3]

On 14 August, 1863, in an action against the forts com-manding the harbour of Kanagawa, the Armstrong R.B.L.s performed badly. Five ships fired 365 rounds from 21 guns and there were 28 accidents. Moreover, shooting was erratic and many shells failed to explode. As a result of this affair the 7-in. guns were withdrawn from service, and by the end of 1864 the use of breech-loaders had ceased in the Navy. Both Services were agreed that the breech-loaders were unsatisfactory (see Chapter VIII), but whereas the Army wanted rifled muzzle-loaders, the Navy had a hankering after the old and well-tried smooth-bore pieces. They were simple and durable; up to 1,500 yards they were as accurate as rifled guns, and most naval officers did not expect to engage enemy ships at any greater range.

Because the War Office was the controlling authority in the supply of ordnance, it was difficult for the Admiralty to follow an independent line. Nevertheless a 100-pr smooth-bore gun, known as a 'Somerset', was designed by the Admiralty and made by Armstrong, and this proved capable of penetrating 5½-in. armour plating at 200 yards. But it was difficult to design a satisfactory mounting for using a gun of this size on board ship. Most ships' guns were still mounted on the old truck carriages, but the Somerset was too heavy for anything so mobile. Eventually a carriage working on slides was designed and 35 Somersets were issued to four ships. But even with these carriages the gun was so difficult to control that production was stopped and the 35 guns were put into store.[4] Woolwich, in the meantime, had designed a heavy smooth-bore piece, and the Admiralty were offered thirteen of these 156-prs, of which five were mounted in the coast defence ship H.M.S. *Sovereign*; but they proved fairly ineffective. The last flirtation with big smooth-bores took place in 1865 when Armstrongs built four monster 600-pr 13-in. guns. But on test their shooting was so erratic that they were never taken into service.[5]

The Navy received its first heavy rifled muzzle-loader in 1865 – a 64-pr designed by Mr Fraser of the Royal Gun Factories which had a rifling cut with an increasing twist, instead of one which was uniform, and intended to take studded projectiles. It was not a success; its shooting was erratic, and the rifling scheme resulted in damage to the gun. Nevertheless it was fifteen years before the Navy obtained a replacement.[6] A much better R.M.L. gun, the 9-in., was produced the following year and the battleship H.M.S. *Bellerophon*, completed in 1866, was the first to be armed with it. Ten of the new guns were installed in the ship mounted on metal carriages with slides.[7]

When H.M.S. *Hercules* was building, the Ordnance Department announced that the 10-in. R.M.L. (the advent of which is recounted in Chapter VIII) would soon be available, and when the ship was completed in 1868 she carried four a side, mounted as broadside guns in a central battery. But the following year was far more important in naval annals, for it saw the completion of H.M.S. *Monarch*, the first battleship to have her main armament mounted in turrets. There were two of these amidships on the centre line and in each of them were two of the Navy's first 12-in. R.M.L. guns, firing a 600-lb shell and sighted to 700 yards. This large piece acquired the nickname of 'Big Will'. The advent of the turret, however, was of far greater interest than that of the guns which it housed.

During the Crimean War, Captain Cowper Coles R.N. designed and built in 1855 an armed raft for coastal operations in the Sea of Azov. As an improvement on this he proposed an armoured raft with a gun in a fixed cupola. From this he developed his idea of a revolving turret, which he patented on 3 March, 1859; and in the same year the Admiralty agreed to finance the manufacture of an experimental turret. Little was done during the next two years, but in 1861 the project was transferred to Woolwich, where a turret was completed and successfully passed the most stringent tests.[8]

Cowper Coles had not been alone in his conception of the turret mounting. In the United States the Swede John Ericsson had built the U.S.S. *Monitor* with a turret, and the ship had acquired fame through its duel with the Confederate warship *Virginia* (ex-*Merrimac*). As a result ships with turret mountings were being built in the United Kingdom for foreign countries. In 1861 Denmark ordered a gunboat, the *Rolf Krake*, which had two two-gun turrets and was of great value in the war against Prussia in 1864. Holland, Italy, Brazil, Prussia, and Russia had all placed orders for turret ships with British firms, and two building for the Confederate States of America (the *Wyvern* and the *Scorpion*) had been seized and purchased by the British Government.[9]

Accordingly, turrets were now tentatively introduced into the Royal Navy. The first ships to have them were coast defence vessels. In 1862 the 131-gun *Royal Sovereign*, newly completed, was selected for cutting down and conversion to turret armament for her main guns, and at the same time a new coast defence turret ship, the *Prince Albert*, was ordered. The conversion of the *Royal Sovereign* was completed in 1864 and she had four turrets, one with twin guns and the others with one gun each. The *Prince Albert* was completed in 1866 with four turrets, each mounting one 9-in. R.M.L. gun.[10]

Captain Cowper Coles, in the meantime, had been urging the construction of a battleship with the main armament in turrets. As a result, two were laid down, the *Monarch* to an Admiralty design and the *Captain* built by Messrs Lairds to embody Coles's ideas. The *Captain*, completed in 1870, also had two 12-in. R.M.L.s in each of two turrets, but she lacked stability and capsized in a heavy sea, going down with Captain Cowper Coles and almost her entire complement.

But the disaster of the *Captain* had nothing to do with turrets as such, and turret-mounting for guns had, of necessity, come to stay. It was still possible to work the existing guns on the broadside, but only just. Bigger guns would have to be mounted on balanced turntables secured by central pivots on the centre line of the ship.[11] Masts and sails now, logically, disappeared, because they did not mix with turrets. The *Devastation*, laid down in 1869, was able to carry heavier 12-in. guns and double the weight of armour. She was followed by the *Thunderer* and the *Dreadnought* which were developments on the same theme. In the *Devastation* the twin 35-ton guns in each turret were loaded and worked by hand. In the *Thunderer*'s forward turret two 38-ton 12-in. guns had Armstrong's hydraulic system, and this was used in both turrets of the *Dreadnought*. In the Armstrong system the turrets were revolved by steam. After firing, the turrets were turned from the firing position until the muzzles were over the loading tubes which led up from the lower deck at an angle of 30° on either side of the middle line, the guns being depressed to the requisite angle by hydraulic power, and telescopic hydraulic rammers forced the 700-lb shells and their cartridges into the muzzles.[12] After being loaded, the turret was rotated towards the target and the guns were fired by electrical control from the conning tower.

The arguments for a return to breech-loading have been given in Chapter IX. As far as the Royal Navy was concerned, the matter became urgent when anxiety was aroused by the news that the French Navy had been equipped with efficient and powerful large-calibre breech-loading guns. In 1879 Armstrong completed some gunboats for China, each armed with two breech-loading guns. The arrival of these at Spithead aroused interest not unmixed with dismay, because their guns were so obviously more powerful and efficient than any mounted in the Royal Navy of equivalent size.

The outcome of these anxieties was the laying down in

1880 of the *Collingwood*, the design of which was influenced by the French battleship *Caiman*. The *Caiman* was laid down at Toulon in 1878, armed with two 16.8-in. breech-loading rifled guns, each mounted in a barbette, one at the bow and the other at the stern.

(The difference between the turret and the barbette is a little difficult to appreciate. The original turret was a generally cylindrical armoured housing, mounted on central pivots for rotation. The barbette, on the other hand, was a stationary armoured tower which enabled the gun to be mounted at a greater elevation than in the ordinary turret. At the top of the tower was a turntable on which the gun or guns were mounted. Later a steel hood was placed over the gun for protection and in due course this became a barbette-turret. Eventually the term 'barbette' was dropped from normal usage, leaving 'turret' as the common designation.)

The *Collingwood* had two 12-in. 43-ton guns at both bow and stern, mounted on barbettes, and only 6-in. guns on the broadsides. She may probably be regarded as the first modern battleship.[13] Five more ships of the *Collingwood* type were laid down in 1882–3 and were known as the 'Admiral' class. Their armament was heavier than that of the *Collingwood*. Four of them (*Anson, Camperdown, Howe*, and *Rodney*) had four 13.5-in., 67-ton B.L. guns disposed in two barbettes. The fifth, the *Benbow*, had two enormous 16.25-in., 110-ton B.L. guns, one in each of two barbettes. This ship had been put out to contract and because it would be some time before Woolwich could deliver the new 13.5s and the Admiralty were anxious to avoid any delay in the delivery of her armament, it was decided to select heavy guns from those immediately available. These were limited to 12-in. and 16.25-in. and the latter was chosen. The barbettes in all these ships were open.

In the next two battleships, the *Sans Pareil* and the *Victoria* (laid down in 1885), a reversion was made to turrets, and the 16.25-in. gun was apparently regarded with sufficient favour for two of them to be mounted in one turret forward. The *Nile* and the *Trafalgar*, laid down in 1886, also had turrets, but the 13.5-in. gun was selected, and there were four of them in two turrets. But the chief interest in these battleships was their secondary armament of Q.F. guns – at first 4.7-in. and later 6-in. They were the first British battleships to have a Q.F. secondary armament.

With the 'Royal Sovereign' class of eight battleships, laid down in 1889–91, a final reversion (with one exception) was made to barbette mountings. The exception, in deference to the views of the First Naval Lord, Admiral Hood, had turrets and was named after him. The others were far better ships and the barbette was adopted for all future battleship construction. All of the class had four 13.5-in. guns as their main armament and sixteen 6-in. Q.F. as their secondary.

In the 'Majestic' class, laid down in 1894–5, a return was made to 12-in. guns and the barbette, instead of being open, had an armoured hood covering for the guns. The battleship armament was now becoming standardized, and until the advent of the 'Dreadnoughts' it consisted of four 12-in. guns, 4.7-in. or 6-in. Q.F. guns as a secondary armament, and a number of 12-prs as an anti-torpedo armament. This was modified, it is true, in the last two classes of 'pre-Dreadnoughts', the 'King Edwards' and 'Lord Nelsons', which had the short-lived aberration of a mixed main armament; for, in addition to the four 12-in., the former had four, the latter no less than ten 9.2-in. guns. However, the 'Lord Nelsons' had no secondary armament.

The *Dreadnought*, completed in 1906, was a revolutionary battleship – so revolutionary, in fact, that on her appearance every other battleship in the world was obsolete. She was the all-big-gun ship with ten 12-in. guns in five barbette-turrets, no secondary armament, and twenty-seven 12-prs as an anti-torpedo defence. Successive 'Dreadnought' classes culminated in the 'Iron Dukes', which were the last to be completed before war broke out. By this time the main armament had become ten 13.5-in. guns disposed in five barbette-turrets along the centre line of the ship and 6-in. Q.F. guns as a dual secondary armament and anti-torpedo defence.

In the 1912 and 1913 programmes main armament guns were larger and fewer. The 'Queen Elizabeth' and 'Revenge' class battleships had eight 15-in. guns and fourteen 6-in. guns. But a school of naval thought was arising which queried the emphasis on big guns and the policy adopted in the construction of battleships. The matter was ventilated in a book by Admiral Sir Reginald Custance entitled *The Ship in the Line of Battle*. The principal doubt was whether slightly thicker armour and speed increased by up to four knots really compensated for a reduction in the number of guns. It was suggested that the ship with the larger number of guns would at any probable range obtain the greater number of hits. An interesting comparison was drawn between two ships under construction in England, one for the Royal Navy and the other for a foreign power. It was suggested that the main respective features were:

	Queen Elizabeth	*Rio de Janeiro*
Displacement	27,500	27,000
Main Armament	8 x 15-in.	14 x 12-in.
Secondary Armament	16 x 6-in.	20 x 6-in.
Speed	25 knots	22 knots
Armour	13-in., 10-in., 9-in., 6-in.	9-in., 7-in., 6-in.

(The *Rio de Janeiro* was taken over as H.M.S. *Agincourt*.) It is doubtful whether these matters were ever resolved to everybody's satisfaction.

Jellicoe, who became Second Sea Lord in 1913, was not happy about the relative powers of British and German battleships. Although German battleships carried guns of smaller calibre than their British contemporaries (*e.g.*, 11-in. instead of 12-in., and 12-in. instead of 13.5-in. or 15-in.), the striking energy of the German projectiles, owing to their much higher muzzle velocity, was little less than that of the corresponding and larger British guns. And saving weight on armament allowed the Germans to provide greater weight of armour.[14]

But apart from guns, ammunition presented a problem. In 1908 Jellicoe became Third Sea Lord and Controller, and he was worried about the effectiveness of shells at long range when they would have an oblique impact. In 1910 he arranged trials and these showed that the shells frequently failed to penetrate the armour of the target ship. He consequently requested the Ordnance Board to produce a better armour-piercing shell.[15] The D.N.O., who worked under Jellicoe, put before the Ordnance Board the Navy's specifications for an armour-piercing shell which would penetrate armour at an oblique angle and burst inside it. The Board, however, approved a design based on normal impact (*i.e.*, approximately at right angles) and argued that it would also be best for oblique impact. The Navy promptly asked for trials with this shell, but the request was turned down on grounds of expense. Jellicoe unfortunately vacated the office of Controller at this time and the matter was not pressed. The approved shell, in fact, was apt to detonate on striking even fairly thin armour at right angles, and before the fuse had operated. The trouble was due primarily to the sensitive lyddite filling of the shell.[16] This appalling defect can be attributed only to criminal stupidity or incompetence on somebody's part.

Having discussed the heavy armament of the Navy, we may now have a glance at the guns mounted on the smaller ships, taking the destroyers as an example.

Robert Whitehead's first torpedo, made in 1866, ushered in a revolution in naval warfare that can hardly have been foreseen. In 1870 the British Government purchased the right to manufacture it. In 1873 Andrew Yarrow mounted a spur torpedo in one of his fast launches and was soon building torpedo boats for several foreign countries. This introduced such a dangerous threat to battleships that a new class of naval vessel soon made its appearance with the primary task of destroying the enemy's torpedo boats – hence its eventual name of torpedo boat destroyer.

The first British torpedo boat destroyers were the A class of 1892, armed with one 12-pr and three (later five) 6-prs. The armament of these first destroyers was retained for the

16. An 8–in howitzer at Wagonlieu in 1916.

17. The crew of a 15–in howitzer with their weapon.
Englebemer Wood, 1916.

18. An enormous French 370-mm (14.8-in) howitzer in the Ravin de la Baraquette, near Foucaucourt, 1916.

19. The British 6-pr anti-tank gun.

various subsequent classes up to E; but for the 'Tribal' or F class of the 1905 programme the 6-prs were abolished and the 12-prs increased to three. But the 'Beagle' or 'G' class of 1908–9 had a much more powerful armament of one 4-in. and three 12-prs. The 4-in. was a 26-cwt B.L. (not Q.F.) equipment firing a 31-lb shell. In the succeeding 'Acorn' or 'H' class (1909–10 programme) the 4-in. guns were increased to two and the 12-prs reduced to two. The 'Acasta' or 'K' class of 1911–12 had three 4-in. pieces and no 12-prs; for it had been decided that there was no point in retaining 12-prs because the 4-in. was so much the better gun. Tests had shown that one shell from the 4-in. gun did more damage than six from a 12-pr. The 'Acastas' were the last destroyers to be armed with B.L. guns, and the 'L' class of the 1912–13 programme had three 4-in. Q.F. equipments. This armament was repeated in the 'M' class, the first of which was launched just before the outbreak of war.[17]

By this time torpedo boats (the original reason for the torpedo boat destroyer) had practically disappeared from the naval scene; their functions could be more effectively performed by the destroyers, which gradually lost the 'torpedo boat' part of their name.

CHAPTER XIII

Balkan Interlude

THE Balkan Wars of 1912–1913 have a particular artillery interest because they were the first major operations in which both sides were armed with Q.F. field equipments. Furthermore, it was the Turkish field guns that saved their army in Thrace from collapse and, consequently, Constantinople from Bulgarian occupation. Otherwise, the course of history might well have been changed at the expense of the Western Powers.

These very interesting campaigns in the Balkans have been largely lost sight of because the much greater First World War broke out before any proper study of them could be made or any comprehensive history written.

The Turkish artillery was armed mostly with Krupp equipments whereas the Balkan allies bought practically all their ordnance from Schneider. The Turkish field gun was the 75-mm. Krupp pattern mentioned in Chapter XI, firing a shell of slightly over 13 lb with a maximum range of 8,700 yards. The Bulgarians, Serbians, and Greeks had the standard Schneider 75-mm., firing a projectile of about the same weight as the Turkish but with a shorter range of 7,665 yards. The Turks had the 120-mm. (4.7-in.) Krupp howitzer but with, as already stated, controlled recoil. The Bulgarians had both Krupp and Schneider 120-mm. howitzers, the former having the normal Krupp rear trunnions. The Schneider 120-mm. with controlled recoil, which had been purchased just before the war, appeared to be the better weapon. It was also the sole field howitzer in the Serbian and Greek armies. The best heavy gun was the new 120-mm. Schneider, but only the Serbians had it. The Greeks and Turks had some 100-mm. and a few

150-mm. Krupp guns.[1] The Bulgarian siege train for the attack of Adrianople consisted of not more than 50 Schneider 120-mm. and Krupp 150-mm. guns, together with a few Schneider equipments of 240 mm.[2] The mountain guns of the opposing powers are given in Chapter XI.

All the armies used both shrapnel and H.E. ammunition, though their heavy pieces fired H.E. only. The Bulgarians had an H.E. shell burst by time fuse, which had a very great morale effect. According to Serbian officers, when they were fighting against the Bulgarians, this shell upset the nerves of entrenched infantry and gunners behind shields much more than did shrapnel. In addition, if it burst at the right height no gun shield could withstand it, and the resulting material destruction was considerable.[3]

Most field batteries had four guns, three batteries formed a group, and three groups made up a regiment of field artillery. In the Bulgarian and Serbian armies there was standard allotment of one field artillery regiment in each division (though in a Serbian 2nd Ban Division there was only one group). Heavy and mountain artillery were army troops.[4] Probably the best field artillery was the Serbian. The officers were keen and highly trained, and battery commanders displayed considerable initiative. In the fighting against the Bulgarians the Serbians used telephone lines to maintain communications between artillery and infantry, with surprisingly good results.[5]

Both Serbian and Bulgarian batteries almost invariably came into action in covered positions and fired by indirect laying, and guns were often entrenched. The Greeks tried at first to do the same, but their gunners were neither trained nor equipped for indirect laying and they soon reverted to siting their guns in the open.[6] The Turks always brought their guns into action in the open and usually sited their batteries either on forward slopes or just behind crests, much as the British artillery had been sited at the start of the Boer War. Like the British, the Turkish gunners suffered heavily, and the shields of captured guns testified to their high casualties. But in spite of their more advanced methods, the shooting of the Bulgarian artillery was not good. They never seemed to get the correct range and they gave very poor support to their infantry during the attack on the Turkish Tchataldja Lines in front of Constantinople.[7]

The Romanians, who came into the Second Balkan War against the Bulgarians at the last moment, were armed with Krupp 75-mm. guns and Krupp 120-mm. and 105-mm. howitzers. In addition, they were the only one of the contesting powers to have a British equipment, the Elswick mountain gun.[8]

The Q.F. gun played its first decisive role in history in the campaign in Thrace between the Turkish and Bulgarian armies in 1912, for it saved Turkey from disaster and perhaps ensured that Constantinople, the present Istanbul, is not now a Russian city.

The Turkish Army was quite unready for war: the reserve divisions and the reservists who made up the bulk of the infantry in the active divisions were quite untrained; equipment and material were deficient; administrative services were inefficient or non-existent; and through ill-conceived reorganisation, morale was low and there was a grave shortage of regimental officers. Only the cavalry and artillery were dependable fighting arms, especially the latter.

The Bulgarian Army, on the other hand, was as ready as, in the opinion of General Savov, the Chief of Staff, it ever could be with the resources available. There was merely the matter of selecting the appropriate time. The Bulgarian soldier began his service with the colours in February and completed it in October of the following year. The standard of unit and formation training was at its best, therefore, in the autumn. By the end of September the harvest was over and the vineyards plucked. This, then, was to be the time when mobilisation should begin.[9]

Thrace was a thinly populated region, underdeveloped, and very poor. Communications were few and bad; only one railway and one metalled road crossed the Turco-Bulgarian frontier and both these followed the Maritsa valley. All other roads were unmetalled tracks across the virgin earth which dissolved into almost impassable mud after heavy rain. The principal rivers, the Maritsa, Tunja, and Arda, were considerable obstacles to military movement; for the Maritsa, below Adrianople, was as wide and deep as the lower reaches of the Thames and the other two were about the same width and depth as the Thames at Windsor. All were liable to flooding and there were only a few bridges.

The Bulgarian General Staff was well acquainted with the distribution of Turkish forces in Thrace; there were four Army Corps with headquarters respectively at Constantinople (I Corps), Tekirdag (II Corps), Kirk Kilissa (III Corps), and Adrianople (IV Corps), each of which in peacetime consisted of three 'Nizam', or active, infantry divisions. On mobilisation these Corps were reinforced both by reservists to bring the Nizam divisions up to war strength and complete Redif, or reserve divisions.

The two natural lines of invasion, the Maritsa and Tunja valleys, were blocked by the first-class fortress of Adrianople. The Bulgarians, therefore, decided to advance east of Adrianople and parallel to the Tunja, leaving the task of surrounding and masking Adrianople to a separate force. Kirk Kilissa, a smaller fortress east of Adrianople would have to be taken.

The Bulgarian field army consisted of three armies, each of three divisions, and two other independent divisions. The Bulgarian division was very strong and about equal in numbers to a Turkish corps, whilst a Bulgarian brigade was equivalent to a Turkish division. On mobilisation the Bulgarian divisional artillery of nine Q.F. four-gun batteries was increased by the addition of six six-gun batteries *à tir accéléré* (i.e., hotted up pre-Q.F. breech-loaders).[10] Some of the heavy artillery was ox-drawn and Sir Reginald Rankin Bt., in his *The Inner History of the Balkan War*, says that he saw two batteries of 6-in. guns in which each gun was drawn by a team of ten oxen.

Mobilisation of the Bulgarian Army began on 30 September and elaborate precautions were taken to prevent news of their concentration areas reaching the Turks in order that they might achieve strategical surprise.

One army was to invest Adrianople whilst the other two were to advance east of Adrianople and defeat the Turkish field army in Thrace.

The Turkish Army of the East, as the field forces in Thrace were designated, concentrated east and south of Adrianople, and behind it the Reserve Army was assembling in areas west of Constantinople. General Izzet Fuad Pasha, posted to command a corps in the Reserve Army, found that in one of his divisions only about a third of the men had done any training. The men's cartridges were twenty years old, they had no cartridge belts or haversacks and had to carry their ammunition in their pockets, and transport was in such a poor state that battalions could carry neither supplies nor baggage. On 18 October, Izzet Fuad saw Aadil Bey, an officer of whom he had a high opinion and who commanded a squadron of horse artillery, supervising the entrainment of his guns and men. He was very unhappy and showed Izzet Fuad a list of all the essential things that he lacked, including even horses capable of pulling his guns and adequate stocks of ammunition.[11]

The III Turkish Corps, under the command of General Mahmoud Mukhtar Pasha, was at Kirk Kilissa and had the task of covering the concentration of the Eastern Army. Like every other Turkish formation, it was seriously deficient of trained men, transport, horses, equipment, food, ammunition, and medical organisation.

Mahmoud Mukhtar arrived at his Corps Headquarters on 17 October, and on the 20th was ordered to take up a position on the right of the line. On his left was to be I Corps, and then in order, II Corps and IV Corps. III Corps had three Nizam infantry divisions, 7, 8, and 9, and one Redif (or Reserve) division, the Afium Karahissar. The 8th Division was very weak with only one infantry regiment, one rifle battalion, one field artillery group, and one mountain artillery group. The 7th Division was much stronger, having three infantry regiments, one field artillery regiment, and three howitzer batteries. The 9th Infantry Division had two infantry regiments, one infantry battalion, one Redif infantry regiment, three Redif infantry battalions, one cavalry regiment, and a field artillery regiment of only two groups. The Afium Karahissar Redif Division was also very weak with six Redif infantry battalions and two batteries of mountain artillery.[12]

Towards 7.0 p.m. on 21 October, Mahmoud Mukhtar received the order for a general offensive from Army H.Q. The 7th Division on the right of his Corps had been reinforced, on account of the importance of its sector, with three howitzer batteries, and three extra batteries of Krupp field guns were allocated to the force assigned to the defence of Kirk Kilissa.

Most of III Corps troops had not yet completed their concentration and infantry and artillery were sited in widely separated localities. Darkness added to the difficulties of organising movement. The infantry of the Karahissar Division were able to march at 2.0 a.m., but they had not got their artillery, and the 9th and 10th Divisions did not start till 6.0 a.m. and 7.0 a.m. respectively. By about midday on the 22nd the advancing troops of the Turkish III Corps were encountering the Bulgarians. At near 3.0 p.m. Mahmoud Mukhtar came across the Afium Karahissar Division starting to retreat, because, apparently, the Divisional Commander had seen in front of him an enemy force of all arms and believed he could not oppose it without artillery. Mahmoud Mukhtar gave him peremptory orders to halt his division and take up a defensive position.

Along the whole of the Corps front the untrained and under-officered Turkish infantry was deploying with great difficulty and in complete disorder. To cover this deployment the Turkish artillery were engaged in a duel with the enemy guns. Mahmoud Mukhtar, assisted by his Staff and divisional commanders, was personally trying to organise a firing line. This, he says, was not his job![13]

At 5.30 a.m. on 23 October, Mahmoud Mukhtar was ordered by Army H.Q. to retire to the west and north-west of Kirk Kilissa because I Corps on his left was in bad trouble. II Corps was not yet up and Bulgarians pressing forward on I Corps' open left flank had caused a panic amongst the infantry, who had fled leaving the guns to fight it out. Two batteries of the 3rd Division on the left of I Corps had only one officer left. Some batteries had fallen into enemy hands, but had been recaptured by a counter-attack by the Ismid Division of IV Corps. Another counter-attack had been unsuccessful because Turkish battalions fought each other under the impression that they were fighting Bulgarians.

Mahmoud Mukhtar decided that he had superior forces to the Bulgarians and that he would not therefore conform to the general withdrawal but would attack. Having issued orders accordingly, he was about to mount with his staff when an officer arrived from the front to say that the Afium Karahissar Division was in flight. Leaving the village at a gallop, Mahmoud Mukhtar was soon amongst infantry flying in confusion and bewildered batteries moving rapidly in retreat. Mahmoud Mukhtar and his staff immediately stopped the batteries, ordering them to turn about and open fire, whether there was a target or not. He says, 'Nothing has more effect on an untrained and nervous soldier than the fire of his own artillery.' With the guns to help them he and his staff, laying about with their swords, stopped the panic. It transpired that the Bulgarians had surprised troops in the front line who were asleep without any protective arrangements having been made. The subsequent panic flight had communicated itself, not only to the remainder of the Karahissar Division, but also to the 9th Division.

Little by little, order was re-established and by 11.0 a.m. all batteries were in action and there was some sort of infantry line. But shortly after this Mahmoud Mukhtar heard that part of the line was again giving way as untrained reservists broke into renewed panic. He and his staff and their horses were so exhausted by their previous efforts it was impossible to intervene again, and Mahmoud Mukhtar accordingly ordered General Hassan Izzed Pasha, commander of the 9th Division, to form a rear guard to cover a retreat to Kirk Kilissa. Soon after this he received a message from the Army Commander saying that the troops were in no condition to stand and that the retreat was to continue to the line Lule Burgas–Viza. Mahmoud Mukhtar ordered the 9th Division to continue to provide a rear guard for this further withdrawal, but he was informed later that a great part of the division had taken advantage of darkness to run away. The Divisional Commander, deserted by his troops, arrived alone at Viza.[14]

Leaving Kirk Kilissa, Mahmoud Mukhtar saw the road to Viza covered with guns, personal weapons and fugitives. The road was largely unmetalled and the continuous rain had resulted in guns and vehicles becoming embedded in the mud. He was horrified by the disaster. Without being pursued by the enemy the troops, vanquished by their own fears and the state of the roads, had fled as if they had suffered an overwhelming defeat and had abandoned a third of all their war material.

At dawn on 25 October Mahmoud Mukhtar formed a detachment of one infantry regiment, two mountain batteries, and a section of cavalry, to go back towards Kirk Kilissa and collect the abandoned guns. The following day the detachment was in contact with the enemy, but, after fighting well for a time, one battalion composed entirely of newly instructed recruits broke and fled, a terrible thunderstorm and torrential rain contributing to the collapse of their morale.

On 27 October Mahmoud Mukhtar set about restoring the situation around Viza and reorganising units. Luckily the Bulgarian advance was incredibly slow. He heard that I, II, and IV Corps had fared even worse than his own. By 28 October he had formed three weak composite divisions, but only eight battalions and three batteries were as yet fit to move. The next day he received a very welcome reinforcement of three batteries from the School of Artillery.

On 29 October Mahmoud Mukhtar's reconstituted Corps was moving into action against the now approaching Bulgarians, its morale stiffened by the amazing personality of its commander. Again, however, part of the line wavered. The artillery, the most dependable arm, was all too scanty and some of the infantry on the left broke and fled crying, 'We will not remain without artillery support'. But on the right of III Corps the Turkish infantry were fighting so stoutly that the Bulgarians in front of them were giving way. By 5.0 p.m. the Bulgarians had had enough and were recoiling in disorder along most of the III Corps front.[15]

The part played by the Turkish artillery in the battle was brilliant. Some nine batteries and 6,000 infantry had taken part in this, the first Turkish success, and the bearing of the Turkish infantry showed their heightened morale. Mahmoud Mukhtar ordered divisional commanders to follow up the retreating Bulgarians with advanced guards until dark.

Following his success, Mahmoud Mukhtar was now placed in command of a Second Army, formed by re-grouping and newly arrived reinforcements. He had every reason to be pleased, for the action of III Corps on the 29th had stopped the flank of the army being turned and pushed south, with the inevitable capture of Constantinople by the Bulgarians as a consequence.

On the 30th, III Corps renewed its attacks. Although some of the infantry reservists again gave way, batteries hurriedly brought up in support of the weaker sectors ensured the continuation of the advance. But though the Second Army was fighting successfully, on the left the First Army had suffered a reverse and artillery ammunition was running low.[16]

By 3 November the whole of the Turkish formations in Thrace were withdrawing to the Tchataldja lines in front of Constantinople. All attacks by the Bulgarians against these entrenched positions were smashed, principally by the Turkish artillery. The event, unfortunately, was too close to the First World War for its lessons to be learned.

This account may appear to show the Turkish infantry in a poor light. It must be emphasised, however, that it was generally the untrained reservists who gave way; raw recruits of any nation, particularly if they lack a proper stiffening of officers and senior N.C.O.s, are of little use in battle. But this short campaign did show that, under a determined commander, the new Q.F. field artillery would provide a powerful framework inside which shaken infantry would stand and even attack.

CHAPTER XIV

The First World War

DURING the First World War the gun was the predominant weapon both on land and at sea. It is true that the torpedo fired by the submarine brought Great Britain near to defeat, but it was under the guns of the battle fleet that the submarine menace was eventually defeated. On land, though the machine gun was a formidable weapon of defence, it was only the gun, whether static or in a tank, that could smash a way through to victory.

The Western Front

Certainly on the Western Front it was a 'gunners' war'. Although operations started with the emphasis on field artillery, it was increasingly the heavy artillery that came to play the major role. For with the long line of trenches stretching from Switzerland to the sea, both sides settled down into opposing fortresses and the fighting, on an enormous scale, soon resembled the classic siege warfare of the eighteenth century. Because it was siege warfare, it was the siege gun which both Allies and Germans thought would open the breach in the ramparts for the final assault.

The British Expeditionary Force sailed for France in 1914 expecting mobile operations: an expectation which was, indeed, more than satisfied for the first few weeks. The artillery element for this period was ample, consisting as it did of five Royal Horse Artillery 13-pr batteries in the Cavalry Division, and in each Infantry Division four Royal Field Artillery brigades of 18-pr guns and 4.5-in. howitzers, and one Royal Garrison Artillery heavy battery of 60-prs.

The Germans, although anticipating mobile warfare, had come prepared for siege operations; for the rapid demolition of fortresses was essential to their plans for a rapid victory. It was their great 420-mm. siege pieces that knocked out the Liège forts and opened the way for their advance through Belgium.

The four British equipments were excellent weapons and the equal of any foreign ordnance of similar type. The 13-prs and 18-prs were much improved during the war by the replacement of the spring recuperator by an air recuperator which, being more powerful, permitted the maximum elevation to be increased to 30°. Initially both these guns were firing only shrapnel, but before the end of 1914 the War Office suggested that some H.E. shell might be useful. G.H.Q., B.E.F. were not at first very enthusiastic, but 1,000 experimental rounds having been supplied, they found the results so good that they asked for more – first on a scale of 25 per cent of all ammunition and subsequently 50 per cent.[1] The 60-pr, a really outstanding gun, had a range in 1914 of 10,500 yards. By altering the shape of the shell this was increased to 12,300 yards. Towards the end of the war a gun and carriage were produced of a new Mark II design which was rather similar to the 18-pr. With this equipment the range was increased to 16,000 yards.[2] Throughout the war all 60-pr guns were drawn by horses, generally heavy Clydesdales.[3]

As the fronts stabilised, more heavy artillery was needed. Some of the heavy batteries which were sent out to France were equipped initially with the 4.7-in. gun, which had been made in 1895 and used in the Boer War. It fired a 45-lb shell about 9,000 yards. In spite of its designation as Q.F., it was a fairly archaic piece, but it had been modernised by fitting a hydraulic buffer and spring recuperator. On firing, the gun slid in the cradle to the rear, drawing with it the piston rod of the buffer and the rods of the springs. The action of these was supplemented by a spring spade attachment to check the recoil of the carriage.[4]

The 1896 30-cwt 6-in. B.L. howitzer has already been mentioned. This fired a shell of about 100 lb and had a range of only 6,500 yards, which was 500 yards less than the 18-pr. On the outbreak of war it had been decided to send four batteries equipped with this piece to France immediately, and about another 50 6-in. howitzers were subsequently collected from various stations and despatched overseas. The mounting was designed to be used either as a travelling carriage, or, with the wheels removed and with a top carriage added, as a howitzer bed. With the travelling carriage, 35 degrees of elevation could be obtained and from 35 to 70 degrees with the bed. To fire the howitzer from its travelling carriage, a platform was specially constructed with wooden beams, and two hours were needed to level the ground and lay the platform. The carriage was then screwed to this platform. The equipment had been modernised by the addition of a hydraulic buffer and spring recuperator, but even so it was somewhat antiquated. It was a heavy equipment and twelve draught horses were used to pull it, until in 1915 they were replaced by mechanical traction.[5] Its despatch overseas was, in fact, delayed over this problem of traction because a specially designed lorry with a four-wheel drive which was being constructed to pull the howitzer and carry the gunners was not ready. Once in position it was a very accurate weapon.

Another tolerably ancient equipment was the 6-in. Mark VII B.L. gun of 1898. This was a coast artillery piece, but in 1914 two siege batteries were improvised, each of which had two of these guns. They were mounted on field carriages, fitted with traction engine wheels, and sent overseas. They fired a 100-lb shell and had a range of 17,700 yards.[6]

The first model of a very famous equipment, the 9.2-in. howitzer, underwent trials in June, 1914. It was appropriately known as 'Mother', for it eventually had 450 offspring. On the road it travelled in three separate parts which were principally the barrel, the carriage and cradle, and the bed, each weighing about 4½ tons and mounted on two four-wheel trucks. The weight of the shell was 290 lb and 'Mother' fired this to a range of 10,000 yards. However, this Mark I was the solitary example, for it was decided that a longer range and some modifications were required. The ensuing Mark II design had a range of 12,700 yards, and this was the pattern for all subsequent 9.2s.[7]

In January, 1915 a new design of 6-in. howitzer weighing 26 cwt was prepared to replace the old 6-in. 30 cwt. It had a range initially of 9,300 yards, but this was later increased to 11,800 yards. It was a most successful equipment and its well-designed hydraulic recuperative system made it independent of a special platform. It became the standard medium howitzer and by the end of the war 1,246 equipments had been issued to the Royal Artillery and another 212 to Allied armies.[8]

A really heavy howitzer being needed, a 12-in. design was prepared in 1915, following as far as possible, in order to save time, the plan for the 9.2-in. This was the Mark I; the Mark II of the same year was intended to fire from railway mountings. The weight of the shell for both types was 750 lb and the range was 11,000 yards. A Mark II model which came out in 1917 had the increased range of 14,000 yards.[9]

In about the middle of 1915, whilst the 9.2-in. and 12-in. howitzers were awaited, a number of surplus 6-in. B.L. guns belonging to the coast defence ports had their muzzles cut off and were then bored out to make improvised 8-in. howitzers. The resulting piece with its cradle was mounted on a special and simple carriage with two enormous traction engine wheels, each weighing a ton and as high as the average man. The piece had a range of 10,700 yards. The shell weighed 200 lb and was loaded by a crane fixed to the carriage. The cordite cartridge was in a silk bag and was loaded separately by hand. For firing a wooden platform was prepared on to which the equipment was hauled, carriage recoil being taken up by four very large skotches placed in front of and behind the carriage wheels. A T-shaped brass tube filled with gunpowder was pushed into the vent at the base of the breech and to this was attached the firing lanyard. On the move the howitzer and its carriage were pulled by a caterpillar tractor.[10]

A number of the Coast Artillery 9.2-in. guns were placed on railway mountings and sent out to France in 1916. Old Mark VI pieces, dating from 1888 were sent out first, but these were followed by the new Mark XIV of 1916 which fired a 380-lb shell to a range of 26,000 yards.

In 1917 a new and much better 8-in. howitzer and carriage was produced, designed on the lines of the 6-in. 26-cwt. howitzer and with a modern recoil and recuperation system. First models had a range of 10,000 yards, but for the later ones this was increased to 12,500 yards.[11]

In 1916 a new Mark XIX 6-in. gun was built with a length of only 35 calibres instead of 45, but with the rather shorter range of 17,000 yards as compared with the 17,700 yards of the old gun. The whole equipment, however, weighed only 10 tons instead of 17.

The biggest guns used in France were the colossal 12-in. and 14-in. pieces with respective weights for the whole equipment of 185 and 246 tons. These were of course on railway mountings.

Tanks

The use of land ordnance by an arm other than the Royal Artillery in the First World War was something of an innovation. The Mark I tanks consisted of two types classified respectively as 'male' and 'female'. The latter was armed with machine guns only, five Vickers and one Hotchkiss, but the former had two Naval 6-pr Q.F. Hotchkiss guns in its sponsons and four Hotchkiss machine guns.[12] This use of mobile artillery had been foreseen in 1877 by Major-General C. B. Brackenbury. As military correspondent of *The Times* he had followed the Russo-Turkish struggle at Plevna and had made three deductions: firstly, that troops under cover were practically impregnable against frontal assault; secondly, that the shorter the range, the greater was the effect of artillery fire; and thirdly, that at close quarters artillery fire practically annihilates. He concluded that artillery should take over the traditional function of mounted troops in the attack and should be

20.   The 3.7–in AA gun.

21. The German 88–mm PAK, an AA gun converted to an
anti-tank role.

22. A troop of 25–prs in action near Bardia. Note the position of guns and limbers.

used as an assault weapon. This should be done, he suggested, by mounting bullet-proof shields on the guns,
which, so protected, should advance continually towards
the enemy.[13]

The choice of the naval 6-pr and its mounting in sponsons
was due to the naval experience of the designers of the
tank; for, as recorded in Chapter IX, this gun had been
adopted by the Royal Navy. But the naval version was
40 calibres long, which was too long for tanks, and it was
soon superseded by a 6-pr designed especially for tanks
which was only 23 calibres long.

Western Battlefields

The Western Front in the First World War was perhaps
particularly noteworthy for the mass of artillery employed
by both sides. It was a far bigger concentration of guns
than had ever been assembled previously or was used in
the Second World War; and with the advent of guided
missiles it is most unlikely that such a vast amount of
ordnance will ever be used again. None of the opposing
countries had been prepared for the enormous expenditure
of ammunition that four years of siege-pattern warfare
entailed, but in the early period Great Britain was far
worse off in this respect than either France or Germany.
And in 1915, to the exasperation of the B.E.F., the gun
ammunition of which they were in sore need was being
sent to the Dardanelles. In May, 1915, before the British
attack in the battle of Frezenberg Ridge near Ypres, there
was four-day bombardment by 433 guns and howitzers
of which 83 were heavy pieces. But this used up so much of
the stock of ammunition that there was not enough left
to give adequate support to the attack, and 4.5-in. howitzers
ran out completely. In addition much of the ammunition
was poor and many shells failed to burst.[14] As a result the
attack was a failure.

For some months after the battle of Festubert in 1915
the siege batteries were reduced to twelve rounds per gun
per day; and for the 6-in. howitzers the only ammunition

available was some 6-in. gun shell from Gibraltar which had been condemned as unserviceable in peacetime. For safety it was fired by a specially long lanyard![15]

The Second Battle of Champagne, primarily a French affair, began on 25 September, 1915, and the rapidly increased scale of artillery support is shown by a notice that Joffre ordered to be made known to every soldier. This said that whereas in the attack near Arras, in the previous May, 300 heavy guns took part, in this coming battle there were to be 2,000 heavy and 3,000 field guns. In Champagne the French had a density of 47 heavy guns per mile; for the complementary operations in Artois they had 35. On the British portion of the Artois front there were only 19 guns per mile.[16] The British operations were known as the Battle of Loos, and not only were there too few guns to give adequate support to the attack, but such guns as there were lacked sufficient ammunition. There was a preliminary four-day bombardment, but the heavy guns had to be limited to 96 rounds per twenty-four hours, or an average of one shell every quarter of an hour.[17]

Nearly a year later, in the Battle of the Somme, on a British front of 18 miles, there were 455 heavy guns and howitzers alone against 400 German heavy pieces. In this sort of warfare the heavy howitzer was the most important weapon, because only its shells were really effective against parapets and deep dug-outs. Field artillery was primarily engaged in wire-cutting.

The procedure for artillery bombardment during an attack, at the start of the Somme operations, was that towards the end of the pre-attack bombardment a heavy barrage was put down on the enemy's front trenches. At zero hour this was lifted to the next trench line in rear and then to subsequent lines in accordance with a laid-down programme. During the phase of the Somme offensive known as the Battle of Flers-Courcelette the creeping barrage was used for the first time and the infantry advanced by limited bounds. For this attack the heavy artillery had been concentrated to provide one gun every twenty-nine yards.

At the Battle of Arras in April, 1917, the scale of heavy artillery in the British Army had increased considerably. There were 963 heavy pieces, or one for every twenty-one yards of front, and the following month at the Battle of Messines there were 2,266 guns of which 756 were heavy, giving one of the latter for every twenty yards.[18]

Nevertheless bombardment on this scale had its disadvantages. The standard practice was to blast the enemy position with an avalanche of fire for days on end, using shells from dumps which it had taken weeks to stock. At the end of this softening-up process the infantry advanced under the protection of more artillery fire. Without the barrage in front of them the infantry could not move and when they reached the extreme range of the guns they had to halt until the artillery could be moved forward. This could be a lengthy process because the artillery had destroyed all the roads over which their guns could move. By the time the guns were in their new positions the enemy had reinforced their front and the process began all over again. Obviously, as Brackenbury had foreseen, a mobile assault artillery was required, and this was now available in the shape of the tank. In June, 1917 Major-General J. F. C. Fuller (then a Lieutenant-Colonel) drew attention to the slight depth of the entrenched fronts in comparison with their length, and pointed out that all that was necessary was to maintain an infantry advance of a few thousand yards. But, he added, gun-fire could no longer do this on account of the increasing depth of entrenchments, and the tank was the only weapon which could. The great tank battle of Cambrai in November, 1917 suggested that Fuller was right; the Battle of Amiens on 8 August, 1918 confirmed that he was.[19]

At the end of 1917, siege batteries were brigaded in groups of four, which usually consisted of three batteries of 6-in. howitzers and one battery of 9.2-in. or 8-in. howitzers. All the 4.7-in. guns and 6-in. 30-cwt howitzers had by this time been withdrawn and their batteries re-equipped either with 60-prs or 6-in. 26-cwt howitzers.[20]

Mountain Artillery

From the massive employment of heavy artillery on the Western Front, it is of interest to turn, at the other end of the artillery scale, to the activities of mountain guns during the First World War. As stated in Chapter X the Indian Mountain Batteries entered the war with the 10-pr gun and its recoil carriage variant, the 2.75-in. Their overseas service included East Africa, the Suez Canal area, the Dardanelles, and Mesopotamia. The first to leave India was the 27th Mountain Battery which sailed for East Africa in August, 1914 with six 10-pr guns. It formed part of the reinforcements which were being sent from India to counter the German threat against the widely extended King's African Rifles. It was soon followed by the 7th Indian Mountain Brigade, consisting of the 21st Kohat Mountain Battery (F.F.) and the 26th Jacob's Mountain Battery, which sailed from Karachi on 17 September, 1914 and landed at Suez to join the force manning the Canal line defences. Their 10-prs were improved for this type of warfare by the addition of shields manufactured in the Suez Canal Company's workshops. On the Canal the batteries went into action several times. In April, 1915 the 7th Indian Mountain Brigade joined the Dardanelles expedition, but the 10-pr proved unsuitable for an area where the nature of the ground and the short ranges at which artillery fire was required demanded howitzers.[21]

In October, 1914 another force arrived from India to take part in the East African operations, the 28th Mountain Battery providing the force artillery. The expedition attempted a landing at Tanga in German East Africa against strong enemy opposition. The assaulting infantry were covered by the two 6-in. and four of the 4.7-in. guns of H.M.S. *Fox* (an old light cruiser, built in 1893) and by the six 10-prs of the mountain battery firing from the deck of the transport *Bharata* (a British India 4,000-ton ship of 1902). The battery commander tried unsuccessfully to observe from the masthead! The battery fired about 150 rounds, which, being mostly unaimed, did little damage.

In addition, the guns were difficult to control, for they were only partially anchored by coal bags, and on the iron deck their recoil was unpredictable. The attack was unsuccessful and the troops had to re-embark. (A curious incident was that both sides were attacked by wild bees; but unfortunately this cannot be attributed to the mountain battery's shells landing amidst their hives!)[22]

The guns of the 27th Mountain Battery, later 7th (Bengal), were fitted with gun shields made in Nairobi, and the 28th Mountain Battery, later 8th (Lahore), also received them when it arrived in British East Africa after the abortive attack on Tanga.[23]

The tsetse fly caused heavy casualties amongst the mules in East Africa. In each mountain battery the experiment was made of replacing the mules in one section by locally recruited porters, but this was not a success. Then the local oxen were tried out as pack animals, but this too was a failure. Nevertheless, in spite of acute transport difficulties the mountain guns were invaluable in this type of warfare. Sections were often detached for minor operations, the guns sometimes being in draught behind oxen with their ammunition carried by porters. In the Uluguru mountains, impassable to wheeled transport, the mountain guns on pack mules came into their own.

In 1917 mountain batteries in East Africa were re-organised as four-gun units. Transport remained as great a difficulty as ever. By the end of March, the 22nd Derajat Mountain Battery (F.F.) had no mules left and was relying entirely on porters, of which it had 445. At the end of August it was re-equipped with 2.75-in. guns in place of the 10-prs. In January, 1918, it received (or acquired!) two of the new 3.7-in. howitzers in exchange for two of its 2.75-in. B.L. guns, and these were soon used with great success against an enemy fort. The 3.7-in. required eight mules to carry it, as compared with six for the 2.75-in., but it was very much the better piece. It was made in two portions, the chase and the breech, which were joined by a junction nut. The piece was 46 inches long and weighed 4 cwt. The carriage had a split trail consisting of two legs, each of which was hinged in the middle so that it could be folded when the equipment was in 'short shaft' draught. Recoil was controlled by means of a cradle, hydraulic buffer, and air recuperator. The ammunition comprised shrapnel, H.E., smoke, and star shell; shell and cartridge were loaded separately. It was 1932 before all the Indian Mountain Artillery got this new equipment.

Indian mountain batteries also took part in the Mesopotamian campaign and formed part of the British forces operating in Iraq. In the latter country they acquired some of the transport mules of South Persia, which impressed them greatly. These big 15-hand animals could carry 240 lb, as compared with the 160 lb of the standard Indian mules.

In June, 1920 all mountain batteries were reduced to an establishment of four equipments.

War at Sea

Throughout the First World War the defeat of the British Battle Fleet would have spelled disaster. It is true that the U-boats brought us near to defeat, but it was through action lasting months. If the Grand Fleet had been decisively defeated, there would have been nothing to stop the German Battle Fleet from cutting all communications to the British Isles immediately. Reduced to its essentials, a battleship was a mobile emplacement for heavy artillery, and it was behind the cover afforded by that artillery that trade and transports moved across the oceans. The proper and efficient use of these big guns was from the start of the war a major preoccupation of Admiral Sir John Jellicoe, the Commander-in-Chief of the Grand Fleet. He was only too well aware of the deficiencies of the British armour-piercing shell and he was worried about the relative powers of the British and German battle cruiser squadrons. He thought that the *Derfflinger* was probably armed with 14-in. guns (though in fact she only had 12-in.), and he wrote to the Admiralty that the German 'gunnery in any action in which they have not been hopelessly inferior has been markedly excellent'.[24] But at the Admiralty, Lord Fisher, writing to Jellicoe on 21 January, 1915, felt that 'the immense superiority of the 13½-in. guns must tell'.[25] But the superiority of the bigger gun depends on its shell being at least as efficient as that of the smaller gun to which it is opposed. Admiral Sir David Beatty, commanding the Battle Cruiser Fleet, was at first more optimistic. Writing to Jellicoe after his battle cruisers had been engaged in the Battle of the Dogger Bank, he reported that the German 11-in. shells were no good, seldom burst, and when they did had a very local effect. He considered the German 12-in. to be 'serious', but not to be compared with the British 13.5-in.[26] But writing later to the First Sea Lord, he said that if the battle cruisers at Jutland had been provided with shell capable of doing what was expected of them the result would have been vastly different and the course of the war changed.[27]

As regards the relative strengths of the British and German battle cruiser squadrons, Beatty, on 3 February, 1916, believed the Germans to have seven fast battle-cruisers, four of them armed with eight 15-in., or possibly 17-in. guns, as compared with four fast British battle cruisers armed with 13.5-in.[28] The intelligence information supplied to Beatty was actually very wide of the mark. The first two German battle cruisers on his list were never built; of the others, three had eight 12-in. guns and two ten 11-in. In fact, the German ships were the better, but not on account of the calibre of their guns or their speed. Three British battle cruisers were lost at Jutland through explosions in their magazines. A German report after the battle said that it was only due to the small explosive charges in British shells that there were no explosions in German ships.[29] Jellicoe, when First Sea Lord in 1917, wrote that if British and German guns of similar calibre were compared, the German gun was considerably superior owing to the higher velocity and heavier projectile; and that a German 11-in. was not greatly inferior in power to a British 12-in. Moreover, the German guns had considerably greater range on account of the higher elevation which their mountings could give.[30]

One result, at any rate, of Jutland was that a new shell was produced in 1917 which would have caused much greater losses in German ships at the battle.

CHAPTER XV

Between the Wars

DURING the years between the two World Wars military expenditure in the United Kingdom was cut to a minimum and there was comparatively little development. To the outsider, the most noteworthy change in the Army was probably the gradual replacement of the horse by mechanical traction. But until the 1930s the guns and howitzers of horse and field artillery were still drawn by six-horse teams in all the armies of the world, and in some countries they were still so hauled in the early part of the Second World War.

As we have seen, mechanical traction of heavy artillery was common in the First World War, and it was inevitable, therefore, that it should later be extended to medium and field artillery. The French Army had, indeed, the first motor-drawn artillery unit because a battery of their 4th Heavy Artillery Regiment was converted from horse traction before the outbreak of the First World War. Initially a motor truck for towing a gun, or such other vehicle as a cable wagon, merely replaced a team of horses and a limber, and it was handled in much the same way. However, conventionally wheeled motor trucks had not nearly the cross-country capacity of horses, though they could move faster and farther on roads. A compromise was at first sought by towing a gun along roads with a truck, which either carried or also towed a small tractor; the latter being used to tow the gun across country. This unwieldy combination was replaced by special vehicles designed for the dual role. The French and German armies used half-tracked vehicles. The British Army had the fully tracked 'Dragon' (i.e., 'drag-gun') and the four-wheeled 'Ant'.[1] Tracks, however, had the disadvantage that they were not so suited for long road movement and the dragons were in due course replaced by six-wheel vehicles. Rubber tyres for the guns logically followed and the 4 ft 8 in. wooden wheels disappeared. By 1939 all British artillery was mechanically drawn.[2]

A next logical step was the mounting of ordnance on self-propelled carriages, and this naturally followed the advent of the gun-armed tank. But in fact the first experiments in self-propelled artillery started even before the First World War. In 1906 Ehrhardt and Krupp both built some 50-mm. and 75-mm. anti-balloon guns, and in 1910 the French Army began trials with a 75-mm. anti-aircraft gun mounted on a de Dion-Bouton chassis. This proved very successful and several more were constructed, all of which had long service; the prototype, indeed, was only retired in 1936. After the outbreak of the First World War the Italians equipped themselves with no less than sixteen four-gun, self-propelled anti-aircraft batteries. All of these guns were of course mounted on four-wheeled vehicles; but their inability to move across country did not matter in anti-aircraft defence.

The first British experimental self-propelled equipment was called a 'Gun Carrying Tank' and it was ready for trial in January, 1917. The components were, roughly, the chassis of a Mark I tank and either a 60-pr gun or a 6-in. howitzer with the carriage wheels removed. Although these pieces could be fired from the tracked mounting, they were normally refitted with their wheels (which were carried on the outside of the vehicle) and fired in the normal fashion.

The French also started experiments in self-propelled artillery on tracked chassis during the First World War and produced a number of prototypes. There were a 220-mm. Schneider gun, a 220-mm. St Chamond howitzer, and various pieces mounted on the Renault F.T. light tank chassis. With remarkable foresight the Inspector-General of Artillery Equipment for the French Army suggested that all field and medium artillery should be self-propelled.

Unfortunately for France, the majority of senior French military opinion was against him and all development of self-propelled artillery came to a stop, except for equipments already on order.

Towards the end of the war the American Army also built a few S.P. guns of various calibres, some mounted on tracked chassis designed by J. Walter Christie, whose tank suspension was later to become famous. But in the United States also military opinion was generally against the idea and development ceased in 1922.

In Great Britain in 1924 Vickers designed and built some S.P. equipments, using a chassis of the same type as that of the firm's medium tanks on which was mounted an 18-pr gun. These self-propelled 18-prs were tested on manoeuvres during the following year. As in France, the Director of Artillery liked them, but again most artillery and senior staff opinion was opposed and no further trials took place after 1930.

Before the outbreak of the Second World War, France alone suddenly appreciated the potential value of the self-propelled gun, and the 1936 defence programme announced the intended formation of five self-propelled artillery battalions. But it was too late, and by the time war broke out progress had not advanced beyond a few experimental equipments.[3]

There had been a reorganisation of the artillery in the British Army in 1924. The distinction that had existed between the Royal Garrison Artillery (Heavy, Coast, and Pack), on the one hand, and the Royal Horse and Royal Field Artillery, on the other, disappeared, and the Royal Artillery was reconstituted as one Regiment. In 1938 there was a further reorganisation. The Regiment was divided into two branches, one being the field artillery and the other comprising coast defence, anti-tank, and anti-aircraft artillery. The old term 'brigade' was replaced by 'regiment', and in this, instead of four six-gun batteries, there were two twelve-gun batteries each of three troops.[4]

The comparative innovation of anti-aircraft guns has not so far been mentioned except in connection with self-propelled artillery. The French Army conducted firing trials against balloons at Bourges in 1908, with the result that the high-angle 75-mm. gun referred to above was ordered. The Krupp S.P. anti-aircraft gun was capable of all-round traverse and elevation up to 75°. The cradle was on rear trunnions to make the breech easily accessible and the motor wagon on which it was mounted had independent drive to each of the four wheels to allow limited movement off the road. Krupp also had a smaller equipment which was horse-drawn and intended for use in roadless country. Ehrhardt's 50-mm. gun was mounted on a fast motor car with the object of chasing a balloon and destroying it by rapid fire at short range. The firm also made a 75-mm. balloon gun on a static mounting.[5]

The first British equipment was an improvised conversion of a 13-pr horse artillery gun in 1914; but this was soon superseded by a 3-in. Q.F. 20-cwt gun on a high-angle mounting, which was either fixed for static defence, or secured to a lorry-towed platform for mobile use.[6] This was used throughout the war and was an excellent weapon.[7]

After the First World War there was a rapid development in the speed and height of aircraft and new anti-aircraft guns with improved appliances and high muzzle velocity were designed to meet the new conditions. These were the 40-mm. Bofors Light Anti-Aircraft and the Q.F. 3.7-in. and Q.F. 4.5-in. Heavy Anti-Aircraft guns.

In the early 1930s a requirement arose for a field gun firing a heavier shell than the 18-pr and to a greater range. A number of trials were carried out with various equipments, with the result that a 3.45-in. piece firing a 25-lb shell was selected. A factor that influenced this decision was the large stock of 18-pr equipments in existence; for the 18-pr carriage was strong enough to stand up to the shock of the shell proposed, and the 18-pr piece could be converted to take it.[8] The new equipment was to be an all-purpose field piece, replacing the 13-pr gun, 18-pr gun, and 4.5-in. howitzer, and having H.E. ammunition only with variable charges.[9] (This decision to do without shrapnel was due to the improved fragmentation of the latest pattern of common H.E. shell.[10]) Owing to the lack of money available for defence at the time, it was not then possible to obtain the new equipment, and the much cheaper conversion of the 18-pr was therefore undertaken. Designs for three other equipments were also produced, but they too were of neccessity shelved. These were for a B.L. 5.5-in. gun-howitzer in replacement of the 6-in. howitzer, a B.L. 4.5-in. gun in replacement of the 60-pr gun (both for medium regiments), and a B.L. 7.2-in. howitzer for heavy regiments in place of the 9.2-in. howitzer. The only new equipment actually issued at this period was a 2-pr Q.F. anti-tank gun.[11]

In 1921 a new tank gun, the 47-mm. 3-pr, was mounted in the Vickers medium tank, which was adopted as the standard for the British Army, and until the 1930s this was the only British tank gun in use. It had a higher muzzle velocity than the 6-pr of the First World War, with consequent flatter trajectory and greater accuracy. It had the additional advantage that more ammunition could be carried in the tank. But it did not have the 'punch' of the 6-pr so that it was less effective when used against strong points or against troops behind cover. However, contemporary opinion held that the tank's gun was only needed against other tanks, and that in the mobile warfare, which was anticipated, the tank's machine gun would be more useful against infantry. In fact, the light tanks of this era were only armed with machine guns.[12]

With the start of British rearmament in the period immediately before the outbreak of the Second World War more modern ordnance at last began to reach the British Army. The prototype of the 25-pr gun-howitzer had undergone trials between 1935 and 1937, and in 1939 it was issued to the Royal Artillery. It had a range of 13,400 yards and soon showed itself to be one of the best field equipments in any army. It was provided with a firing platform which was very similar to one that had been designed for the 18-pr Mark IV carriage. The platform was circular and the spade was anchored to the middle of it by connecting rods. The gun could thus be traversed rapidly about the centre of the platform to face any direction – an invaluable feature for engaging tanks. The platform was slung under the trail for transport. The gun was later fitted with a muzzle brake.[13] Although this equipment was designated Q.F., fixed ammunition was abandoned in order that the charge could be adjusted to the type of target engaged. There were three different charges and a separate supercharge for the 20-lb armour-piercing projectile. The variety of charges allowed for a steep angle of descent when required, and it was this attribute which caused the gun to be designated a 'gun-howitzer'.[14] The gun team comprised six numbers: No. 1 commander, No. 2 breech, No. 3 layer, No. 4 loader, Nos. 5 & 6 ammunition.

The prototype of the 5.5-in. (140-mm.) medium gun-howitzer, with a calibre length of 30, was produced in 1938 and trials with it were carried out in 1938 and 1939. Issue to the Royal Artillery had not started when the Second World War broke out. The cradle had rear trunnions to allow for recoil, and, since this made it muzzle-heavy, two spring equilibrators to compensate for this were fitted, one on each side of the piece, looking like vertical horns.[15] The range of the 5.5-in. firing a 100-lb shell, was 16,000 yards and it had five charges. The 4.5-in. gun, which came out about the same time, was a relined 60-pr (though later marks were new pieces). It was 41 calibres long, fired a 55-lb shell, had a maximum range of 20,500 yards and had three charges. It was mounted on the same carriage as the 5.5-in. also having the rear trunnion cradle and equilibrators. This carriage had a split trail which allowed a 60° traverse, and both could be depressed and elevated between −5° and +45°.[16]

The 7.2-in. B.L. howitzer did not arrive till 1941, but since it was designed before the war its particulars can justifiably be given here. It fired a 200-lb shell over a maximum range of 13,200 yards, using the top charge of its four. It was mounted on a 8-in. howitzer carriage.[17]

Coast armament, when the First World War came to an end, depended primarily on the B.L. 9.2-in. and 6-in. guns on a barbette mounting which had been introduced early in the twentieth century. The 9.2-in. had a range of 15,000 yards with a 380-lb shell, and the 6-in. fired a 100-lb shell over a range of 11,000 yards. These ranges had by this time become too short, owing to the development in naval armament. In the 1920s, therefore, the mountings were altered to increase the maximum elevation from 15° to 35°. In addition the muzzle velocity of both equipments was increased and shells designed of a better ballistic shape. The result was to increase the range of the 9.2-in. to 30,000 yards and the range of the 6-in. to 25,000 yards. Thus altered, these grand guns sufficed for coastal defence until they were abolished in 1956. For defence against light craft, there was a new Q.F. 6-pr twin-barrelled gun with a rate of fire of 60 rounds a minute. It could also be used against aircraft and for this secondary role it had a maximum elevation of 70°.[18]

In the years immediately before the Second World War there was considerable development in the tank guns of the major powers. The main problem was to provide guns sufficiently powerful to penetrate the opposing armour, but no larger than this, so that the maximum amount of ammunition could be carried in the tank. The standard British tank gun at the outbreak of war was a 2-pr (40-mm.) with a calibre length of 52 and a muzzle velocity of 2,650 ft per sec. This was subsequently much criticised as being too small. But there was no apparent reason for it to be any larger. The 2-pr could penetrate 60 mm. of armour plate at a distance of 500 yards, and no German tank at this time had that amount of protection. Indeed the Germans themselves could hardly claim a formidable tank armament: the Pz.II had a 20-mm. gun, the Pz.III a 37-mm. (equal to a 2-pr) and the Pz.IV a low velocity 75-mm. A tank could carry 150 rounds of 2-pr ammunition, the figure which was regarded by the War Office as the desirable minimum.[19]

Between the wars mountain artillery was probably in action more than any other type owing to the numerous campaigns on the North-West Frontier of India. An important improvement in pack saddlery was made in the 1920s, roughly coinciding with the general issue of the

3.7-in. howitzer. For many years the heavier loads had been strapped on top of a saddle-tree which was specially constructed to take a particular load, e.g., breech, muzzle, carriage, etc. The saddle-tree itself rested on two panels stuffed with horsehair to fit the back of a mule. For ammunition and stores, carried in boxes, there was a saddle with two hooks on each side of the mule on which the boxes were slung. Relief mules were provided for the eight loads constituting the 3.7-in. equipment, and because each load had a different saddle this entailed the provision of eight mules. The improved pack saddle invented by Lieutenant-Colonel G. P. MacClellan could accept alternative loads, thus reducing the number of relief mules, with their drivers, to four per gun.[20]

The cartridge of the 3.7-in. howitzer consisted of a brass case with percussion primer and five ordinary charges, of which the fifth was the full charge. The maximum range with shrapnel, using the full charge, was 4,500 yards. The charge was made up of a core of cordite, around which were detachable rings, making five propellants weighing from 3 oz. for the first charge to 9 oz. for the fifth. In later years a super-charge was added which increased the maximum range with shrapnel to 7,000 yards.

Naval Guns

The two battleships of the 1922 programme, *Nelson* and *Rodney*, were notable for carrying the biggest guns ever mounted in a British warship. The armament consisted of nine 16-in., twelve 6-in., and six 4.7 A.A. guns. The 16-in. guns were mounted in three turrets and were the first British warships to have triple mountings. The main armament in the three turrets was concentrated forward. With this arrangement, however, there was a heavy blast in the tower from gunfire abaft the beam; and Oscar Parkes, in *British Battleships*, says that during the Bismarck action in 1941 blast from the *Rodney*'s guns blew the steel helmet from the Captain's head, and it hit and knocked out a signalman standing some feet away from him.

On 31 December, 1936 the Washington Treaty, which had limited the size and number of warships, came to an end. This raised a problem as to the calibre of the guns to be mounted in the next British battleships. Five battleships had been ordered because this was the largest number for which big guns could be manufactured from existing resources. An agreement had been drawn up in 1935 which limited the calibre of guns to 14-in., and a British design for such a gun had been prepared. But Japan had not ratified this agreement and the United States had agreed to wait till December, 1936 when, if there had still been no Japanese ratification, they would build 16-in. guns. But it was essential that guns for the new British battleships should be ordered by June, 1936 if construction of the ships was not to be delayed. If 16-in. guns were ordered it would in any case be a year before they were ready because the designs for their mountings had not yet been made. It was eventually decided to have ten 14-in. guns, mounted in two quadruple and one twin turret; and these were the first quadruple mountings in a British ship. The secondary battery was an innovation, for it consisted of sixteen dual purpose surface and A.A. 5.25-in. guns in eight turrets. The 14-in. guns had a rate of fire of two rounds per gun per minute, a range of 36,000 yards, and a penetrating power of 13 inches of armour at 15,000 yards. The 5.25-in. guns were fully automatic, with a rate of fire of 18 rounds per minute and a range of 22,500 yards.[21]

CHAPTER XVI

The Second World War

Production

Only belatedly did re-equipment of the Services start, and though much was wanted the resources available were all too few. In 1936 the War Office's sources of gun manufacture were the Royal Arsenal at Woolwich, Vickers-Armstrong, and William Beadmore & Co., and their joint capacity was limited. But authorised orders beyond this capacity were too small and extended over too short a period to interest other industrial firms. The Director-General of Munitions Production stressed repeatedly that, if the equipment was to be obtained, orders must be larger and extended over a long term. Requirements for the new 25-pr gun-howitzers and 2-pr anti-tank guns alone could not be met without additional manufacture. In October 1936 a group of firms agreed to forge gun barrels, and the Director-General of Munitions Production, on his own initiative, ordered 500, mostly for the 3.7-in. anti-aircraft equipment. This, however, was in excess of the available authority, and the Treasury objected. Eventually the order was given official approval, but with the proviso that no forgings for 25-prs were to be included.[1] That the D.G.M.P. should have been hampered by such financial obstruction is almost incredible.

At the time of the Munich crisis in 1938 the supply of 352 3.7-in. anti-aircraft guns had been approved but only 44 were as yet available. The bulk of the anti-aircraft artillery, therefore, was supplied by 298 of the old 3-in. guns of 1914, which were quite unsuitable for use against the aircraft of the time. When the crisis had subsided, the

178

War Office announced that it wanted 2,226 field and anti-aircraft guns. This was large by peacetime standards, but in April, 1940, just before the German attack on the Western front, a British army on the Continent of 32 divisions was being planned, and for this were needed 12,677 field and anti-aircraft guns and 13,561 tank and anti-tank guns.[2]

The loss of the B.E.F.'s equipment at the evacuation of Dunkirk in 1940 resulted in the arrival of an unusual piece of ordnance for the British Army, for in response to the Prime Minister's appeal the American Government despatched a large consignment of weapons which included 895 75-mm. guns, together with their ammunition.[3] Many of these guns were allotted to the divisions deployed to meet the expected invasion and there was a weird assortment of improvised transport to move them about their field artillery duties: '*portée*' was the word used at Headquarters II Corps in East Anglia to describe the method of moving them.

In the spring of 1941 came the first demand for an anti-tank gun of larger calibre than the 2-pr. This was occasioned by Rommel's first advance in the Western Desert, and Headquarters Middle East Command wanted large numbers of anti-tank guns which should be not only bigger but of a more advanced design. There was, indeed, something of a crisis caused by the shortage not only of anti-tank guns, but also of tank guns. Although the 2-pr had been adequate at the start of the war, no design had been undertaken in the decade before the war for its ultimate and inevitable replacement. The 2-pr did well in the first Libyan campaign and was better than the German 37-mm. It was known in the first half of 1940 that the Germans were developing a new 50-mm. anti-tank gun and the War Office increased the armour specification to meet it, but the next British gun, the 6-pr, which had been conceived before the war, was not yet ready. It had been discussed in the War Office in April, 1938, but no other action was taken. Over a year later, in June, 1939, a provisional specification was tabled by the Deputy Chief of the Imperial General Staff and a design to meet it was prepared by the Director of Artillery.

As a result a 6-pr anti-tank equipment was ready for trial early in 1940. On 10 June, after the evacuation from France of the B.E.F., the Ministry of Supply asked the War Office to agree to an order for 400 6-pr guns. The War Office, however, was now worried. The 2-pr was still in massive demand and there appeared to be a danger that production of the 6-pr would slow down supply of the 2-pr, which was already too slow to meet requirements. (They had in fact been told that the production of 100 6-prs a year would entail a yearly drop of 600 2-prs.) After discussion with the Ministry of Supply it was agreed to order fifty pilot models in order to make a start in production. The Ministry of Supply now got worried in its turn, in case this very small order should endanger future mass production. They therefore persuaded the War Office to increase the order to 500. In February, 1941 the matter was discussed by the Defence Committee (Supply), when it was decided that a reduction in the manufacture of 2-prs could not be afforded and that the production of 6-prs must be met from new capacity. This, however, was what the Ministry of Supply had itself already arranged. Nevertheless, the guns were inevitably turned out all too slowly. The first thirty-two appeared in November, 1941, but then the rate of production rose rapidly every month until, in May, 1942, 1,517 6-pr guns reached the Army. But their installation in the tanks did not start till the spring of 1942, and it was not till the following autumn that 6-pr anti-tank equipment reached the Eighth Army. By 1944 the 6-pr in its turn had been practically superseded by the 17-pr tank and anti-tank gun and the 75-mm. tank gun.[4]

Tank and Anti-Tank

If, from the Army's point of view, the First World War had been a 'gunners' war', so was the Second, but in an entirely different way. Siege warfare and the massive deployment of heavy artillery had gone – probably for ever – to be replaced by artillery in an assault role (Colonel Brackenbury's conception) as the main weapon; for in most operations decisive action centred round the gun in the tank and the anti-tank gun, or field gun in an anti-tank role. At sea, however, it was soon apparent that the big gun in the battleship was no longer the supreme symbol of maritime power. The sinking of the *Prince of Wales* and the *Repulse*, and the great carrier battles in the Pacific showed that tactical supremacy had passed from the gun to the aeroplane, and to its prolongation, the bomb and the torpedo.

The most interesting development in ordnance was in the tank and anti-tank field, for competition between armour and gun, and between the ranges of opposing tank armaments led to continual new design and construction. At the start of the war the British Army had no reason to be dissatisfied with its 2-pr tank gun. It is true that the heaviest German tank, the Pz IV, was armed with a 75-mm. gun, but it was a short-barrelled weapon with a low muzzle velocity and poor powers of armour penetration. It could out-range the British 2-pr, but the latter at its chosen distance was the more effective armour-piercing weapon. The 50-mm., which replaced the 37-mm. as the armament of the Pz III tank, had much the same range and penetration as the 2-pr and was not much of an advance on the 37-mm. In anti-tank guns, however, the Germans were far better off. The mobile gun, which was used offensively, was the long-barrelled Pak 50-mm., which was very much superior to the tank 50-mm., and also, therefore, to the 2-pr. As a defensive anti-tank gun, Rommel used the 88-mm. anti-aircraft gun, and in this improvised role it was one of the most successful anti-tank weapons of the war.[5] In fact, the British 3.7-in. anti-aircraft gun was even better than the 88-mm. as an anti-tank weapon and there were three times as many in the Middle East theatre of war, but for some reason its use was only allowed belatedly and then in very limited numbers. The British Army had, of course, the excellent 25-pr gun-howitzer, which was provided with anti-tank ammunition in North Africa and fitted with a double baffle muzzle brake to reduce the recoil from this ammunition.[6]

In 1941 American Stuart tanks started to arrive in the
Middle East. They were armed with a 37-mm. gun which,
though of slightly smaller calibre, had a special 'Armour
Piercing Cap and Ballistic Cap' (APCBC) ammunition
which made it more effective against armour than the 2-pr.
The Germans were using a somewhat similar 'Armour
Piercing Composite Rigid' (APCR) ammunition which
had a hard tungsten core enclosed in a soft metal jacket.
Because it was lighter than the normal projectile it had a
higher muzzle velocity, and penetration was easier because
it was only the small diameter core which had to be driven
through the armour plate. At short and medium ranges it
was very effective, but its velocity fell off so rapidly that at
long ranges it was probably inferior in effect and less
accurate than normal ammunition. In 1942 some of the
Pz III tanks were armed with the long 50-mm. and carried
a proportion of APCR ammunition.[7]

The American Grant tank, with which some regiments
in the Middle East were equipped in early 1942, was armed
with both the 37-mm. and a 75-mm. gun. This latter was
much better than the German short 75-mm., even though
it had only a medium muzzle velocity and no capped
ammunition. However, the Grant was badly handicapped
by the position of its 75-mm., which was mounted in a
sponson on the right-hand side of the hull. As a result it
had only a limited traverse and could not be used at all
when the tank was 'hull-down'.[8]

The British 6-pr (57-mm.) anti-tank gun, which was
being received by units in the Middle East in 1942, was a
very good weapon – even better than the excellent German
50-mm. Pak 38. Before the disastrous battle of Gazala
sixty 3.7-in. anti-aircraft guns had been modified for anti-
tank use, but only twelve had reached the army in the
field, and the performance of these was poor, owing to
defective sights.[9]

By the battle of Alamein the American Sherman tank had
been issued to armoured units of the Eighth Army, with a
75-mm. gun of improved type mounted in a turret. It was
outclassed, however, by the new long 75-mm. with which

some of the Pz IV tanks (though as yet fortunately only a few) had been equipped. The British Crusader tank was now armed with the 6-pr, but it also was inferior to the new German gun. This was realised at the War Office, where planning a 17-pr gun to replace the 6-pr had started at the beginning of 1942.[10]

17-pr anti-tank guns reached the Eighth Army in time for the final stages of the campaign in North Africa; and they were needed, for the Germans now had their formidable Tiger tank which was armed with a 88-mm. gun. It had first been used on the Russian front in August, 1942 as the reply to the Russian KV heavy tank.[11]

Before the German attack the Russians had the enormous number of over 22,000 tanks, mostly T-26s with 45-mm. guns and BTs with 76.3-mm. guns; but the issue of two new tanks had started, the medium T-34 and the heavy KV, both armed with the 76.2-mm. gun. Of these the T-34 became the most numerous of Russian tanks. It inspired another new German tank, the Panther, which was armed with a new 75-mm. gun, 70 calibres long and with a very high muzzle velocity. By the winter of 1943–4 the Russians had an improved model of the T-34 which had a long 85-mm. gun, adapted from their 1939 anti-tank gun and comparable with the original German 88-mm. The new turret and gun were also fitted to some of the KV tanks, and, thus altered, they were designated KV-85. As a heavy tank, however, this was soon superseded by the JS-1 armed with a 122-mm. gun. In the later stages of the war the Germans and Russians were making very similar tactical use of their heavy and medium tanks; in an attack against strong defences the medium tanks usually led whilst the heavy tanks supported them from behind with their big guns.[12] The heavy tanks were, in fact, normally used as self-propelled artillery, though able to turn to a proper tank role if required.

For the invasion of Normandy in June, 1944 there was a new British tank, the Cromwell. The earliest marks were armed with 6-pr; but it was soon apparent that this gun was too light, and later marks had a medium velocity

75-mm., which was, however, still much inferior to the latest German 75-mm. tank gun. Two marks of the Cromwell were armed with 95-mm. howitzers for close support, and were therefore really meant to function as self-propelled ordnance.

Experience in North Africa had shown that the 75-mm. gun was fast becoming outmoded. A 17-pr gun had been mounted in some Shermans and Cromwells (which thus modified were known as Fireflies and Challengers respectively). They were followed by a new tank, the Comet, which had a 17-pr modified with a shorter barrel and rather lower muzzle velocity and referred to as a 77-mm. In action this proved to be a good and accurate gun.

Self-Propelled Ordnance

Self-propelled ordnance was developed quite considerably during the Second World War. Probably the initial impetus came from the desire to have an anti-tank gun which would be mobile enough to be deployed rapidly against tanks.

The first British attempt was the somewhat improvised mounting of a 25-pr on the Valentine tank chassis. But by the battle of Alamein an American equipment had reached the Eighth Army in some numbers. This was the 105-mm. 'Howitzer Motor Carriage M7' which incorporated the M3 medium tank chassis. Using this as a model, the British Sexton was produced, which was a self-propelled equipment consisting of the 25-pr gun-howitzer on the Canadian Ram medium tank chassis. In a more exclusively anti-tank role, the 17-pr was combined with the Valentine tank chassis to form the Archer self-propelled gun.[13]

The Germans started using self-propelled artillery a good deal earlier. In 1940 they made some S.P. anti-tank guns by mounting a Czech 47-mm. on a Pz I tank chassis. This was followed by the manufacture and use of a considerable number of S.P. anti-tank guns both in Russia and the Western Desert. The earlier ones were all improvised and included several different types of gun and chassis. There were also improvised S.P. howitzers for the mobile support of armoured formations.[14] But the greatest German development was in assault guns, which they started making at the beginning of the war. The first type was a low velocity 75-mm. gun (the same as the original Pz IV weapon) on a Pz III chassis. In 1942 this gun was replaced on the same chassis by the improved 75-mm. The success of this combination led to the appearance in 1944–5 of the Panzerjäger which was a combined assault and anti-tank self-propelled gun. Equipments of this nature included 75-mm., 88-mm. and 128-mm. They were so successful that before the end of the war the Germans had more Panzerjäger guns than tanks.[15] After experiencing the value of the German assault guns, the Russians were by 1943 producing a range of assault guns themselves, based on the chassis of the T-34 medium and KV heavy tanks. They were generally used for direct fire and often provided a base of manoeuvre for the tank units.[16]

Mountain Artillery

The Second World War brought motor transport to the mountain artillery. Some units were fully mechanised, others were partially mechanised, retaining mules for the 'gun line'. Mules, however, became increasingly difficult to get. All the 3.7-in. howitzers were fitted with wheels having standard rubber motor tyres. The axles, however, could not withstand the extra strain, and from the end of 1944 special nickel steel axles with ball-bearing hubs were provided. A special fitting was made to the saddle for the carriage of these wheels. Mechanical transport consisted mainly of 'jeeps' and trucks, though all sorts of improvisations were made in the early days of the war. The 22nd Mountain Battery was the first Indian artillery unit to go overseas and it disembarked at Mombasa on 10 September 1939, unaccompanied by mules and drivers. It was immediately provided with mechanical transport, but instead of being towed, the 3.7-in. howitzers were carried in two-ton lorries which had been converted for the purpose, with ramps to run the equipments on and off. There was an additional two-ton lorry for each howitzer to carry 120 rounds of ammunition. The detachment was divided between the two lorries.

In 1944 an Indian Mountain Regiment consisted of three batteries each of four 3.7-in. howitzers. Each battery had as transport twelve horses, ninety-four 'ordnance' mules, eleven 'equipment' mules, thirteen R.I.A.S.C. mules attached, nine jeeps, two 15-cwt trucks, two tractors, two bicycles, two motor cycles, one water truck, and one 3-ton lorry.[17]

Naval Guns

The Second World War saw the end of the heavy gun in the Royal Navy; and, effectively, in all other navies, though it has lingered on in the U.S. Navy for coastal bombardment. At the beginning of the Second World War it was not known whether or not aircraft could sink battleships. The question was all too dramatically answered: of the 28 capital ships sunk during the war 16 were sunk by aircraft, 8 were disposed of by the heavy guns of other capital ships, and 3 were torpedoed by submarines. Aircraft, without any assistance, sank five of the very latest battleships: the *Prince of Wales*, the *Roma*, the *Tirpitz*, the *Mushashi* and the *Yamato*.[18] Right at the end of the war H.M.S. *King George V* took part in the bombardment of Tokyo and other parts of Honshu, the main island of Japan.[19] This was the last occasion on which British naval big guns were to fire.

The last of the British battleships, H.M.S. *Vanguard*, was laid down in 1941. She was armed with eight 15-in. guns, and these guns have a special place in naval history, for the ship was designed for the guns rather than the reverse. In December, 1939 the then Director of Naval Construction suggested that the 15-in. guns from the *Courageous* and the *Glorious* might be used for a new battleship which could thereby be built quickly because the design and construction of gun mountings took up more time than any other aspect of battleship construction. The idea was accepted. The *Courageous* and the *Glorious* had been completed in 1917 as light battle cruisers and armed with four 15-in. guns. In 1924 both were taken in hand for conversion to aircraft carriers and their guns were removed. *Courageous* was sunk by a submarine in 1939 and *Glorious* by the *Scharnhorst* and the *Gneisenau* in 1940. The guns were used in action in 1917 by their original owners, but never subsequently.

But if the big gun was destined to disappear, the smaller gun has as yet a major part to play. As compared with the battleship in its heyday, the destroyer was at the opposite end of the naval scale. Originally termed a torpedo boat destroyer, it was, as the name implies, intended to defend the battleships against the menace of attack by the small torpedo boat. Its subsequent growth has been astonishing. First it eliminated and took over the functions of the torpedo boat; next it gradually assumed duties which had previously been allotted to cruisers; then – its final conventional form – it was, save the name, a cruiser; and now it is, in fact, a battleship, for it is capable of engaging any other surface warship and could sink with ease the most powerful battleships of the Second World War. But throughout its history its gun armament has been remarkably consistent.

The development of destroyer armament up to the start of the First World War has been given in Chapter XII. In the first destroyers to be built after the outbreak of war, the V and W classes of 1916–17, the number of 4-in. guns was increased to four, and there was also a 3-in. H.A. (i.e. high angle) gun for anti-aircraft work; but this latter was missing from the S class of 1917–18 and the number of 4-in. was reduced to three. The reason for this reduction was that certain disadvantages had been experienced with the Vs and Ws on account of their heavy armament, and reports indicated that not many German boats were carrying the increased armament that the Vs and Ws had been designed to meet.

In 1918–19 modified Vs and Ws introduced another new gun to destroyers, the 4.7 B.L. They had four of these and one 3-in. H.A. gun, making them easily the most heavily armed destroyers yet (apart from 'leaders'). The 4.7 fired a 50-lb shell which had a much greater hitting power than the 31-lb shell of the 4-in. On the other hand, being a breech-loader with separate charges, it was slower and in bad weather care had to be taken that the charge did not become wet.[20]

The first major post-war design was the 'Acasta' class of 1927. They had four of the new 4.7-in. Q.F. guns, and this was to become a standard destroyer armament, with or without the addition of a 3-in. H.A. gun. When it came to the 'Defender' class of 1930, it was suggested that they should be armed with 5.1-in. guns, because of the heavier or larger number of guns which were being installed in foreign destroyers. The latest French destroyers had 5.1-in. guns, whereas those of Italy and Japan had six 4.7-in. A further argument was that experience of the First World War showed that destroyers used guns more often than torpedoes. Nevertheless, it was decided that, whatever the number of guns mounted, it would be bad to risk a destroyer armament race with the Japanese by increasing the calibre of the guns.

The 'Tribal' class of the 1935–6 programme had the heavy armament of eight 4.7-in. guns in twin L.A./H.A. mountings with a maximum elevation of 40° for long-range A.A. fire. The twin mounting was new, but had been tried out in the *Hereward* of the 1934 'Hero' class. In subsequent classes the twin mountings were retained but the number of 4.7-in. guns was reduced to six, and two of these five pre-Second World War classes had one 4-in. Q.F. H.A. gun as well.

When the Second World War started large numbers of destroyers were required for escort and patrol – vessels which did not need the armament of the fleet destroyers. Successive batches were built in a series of fourteen 'emergency flotillas'. The first, the 'Oribi' class, had four 4.7-in. Q.F. guns and one 4-in. Q.F. H.A. gun. The 4.7s had a maximum elevation of 50°, and experience soon showed that this was quite inadequate to deal with either high-level bombing or dive bombers. There were difficulties in getting a higher elevation for 4.7-in. guns and the succeeding 'Paladin' class had four 4-in. guns on a 80° H.A. mounting. However, this entailed considerable sacrifice in shell power and immediately succeeding classes reverted to four 4.7s and a single H.A. 4-in. In some still later classes the elevation of the 4.7-in. gun was increased to 55° and the H.A. 4-in. omitted. But in H.M.S. *Savage* of the 'Savage' class a new 4.5-in. gun was introduced. The ship had four of these, two in a twin mounting with 80° elevation, and two in single 55° mountings. The next few classes had the 4.7-in. (presumably because there were not as yet enough of the new gun), but the 'Zambesi' class had four 4.5-in. guns in 55° mountings, and the 4.5-in. now became the standard destroyer armament.

The 'Battle' class fleet destroyers of 1942 had four 4.5-in. Q.F. guns in twin 80° mountings and a 4-in. in a H.A./L.A. mounting; but in a later 1943 batch this last was replaced by another 4.5-in. in a 55° mounting. The last 'conventional' destroyers, the 'Daring' class of 1944, had six 4.5-in. guns in twin 80° mountings.[21]

CHAPTER XVII

Ordnance Twilight

IT IS perhaps a curious aspect of artillery equipment that there has been comparatively little development in ordnance and its mounting since the introduction of the long-recoil carriage. On land the greatest advance has been in the increased mobility of pieces of ordnance, whether mounted in tanks, used as assault guns, or providing close support on self-propelled chassis. But in certain respects there has been a recession, for the big guns at sea, in the siege train, and in the coast defences, have gone, and in many spheres ordnance is merely auxiliary to its ancient rival the rocket. It may be, indeed, that rockets are the sole artillery of the future, for further development of ordnance seems limited, even though the advent of the recoil-less gun has introduced a new feature.

The modern artillery equipment of the British Army is partly of British and partly of foreign design. The most noteworthy change is the replacement of the 25-pr towed gun-howitzer by the self-propelled Abbot 105-mm. gun. The chassis of this is standard for a range of light-tracked vehicles and was developed to War Office specifications from 1958. The drive is provided by the Rolls-Royce K.60 multi-fuel engine. With this chassis, the Abbot is air-portable and amphibious. Fully closed down, it protects its crew of four against nuclear flash and fall-out. With collapsible side screens fitted, it can swim under its own power.[1] The Abbot, then, is very mobile and should meet the intentions of its design to provide close support in the most fast-moving operations. The barrel, like many tank guns, has a fume extractor and also a double-baffle muzzle brake. The mounting permits 360° of traverse and 70°

elevation. The gun is designated Q.F. and the ammunition is semi-fixed. The weight of the shell is 36 lb 6 oz, and there are two brass-capped cartridges containing two separate charges – a super-charge cartridge, and a normal cartridge containing charges one to five. Types of shell include H.E. and the anti-tank HESH (High Explosive Squash Head). The maximum range is about 17,000 yards.[2]

The earliest self-propelled equipment in the British Army was the American M-44 155-mm. howitzer which was acquired in 1958.[3] Its original purpose was for the medium artillery support of armoured brigade groups in the British Army of the Rhine in replacement of the 5.5-in. gun-howitzer. The tracked carriage was specially designed for the equipment and it had a very satisfactory performance. But the top of the carriage was not closed so that the crew had no protection from nuclear weapons or air attack.[4]

The replacement of the M-44 is the later American M-109 howitzer equipment, which has been supplied to several NATO armies and is a self-propelled version of the standard U.S. 155-mm. howitzer, differing externally from the latter in its prominent muzzle brake. It fires a 97-lb shell and has a range of 16,000 yards. Ammunition is separately loaded and the charge is bagged.[5]

The heaviest British ordnance is also American – the M-107 175-mm. self-propelled gun, mounted on a tracked chassis. Its appearance is distinctive on account of its very long barrel, which enables it to reach the very considerable range of 35,760 yards. The weight of the shell so fired is 147 lb, and ammunition is separately loaded with a bagged charge. The maximum elevation is 65° and traverse 60°. Before firing, two spades, one at each side of the rear of the chassis, are lowered and dug into the ground by a hydraulic mechanism.[6]

A very useful little equipment is the 105-mm. pack howitzer of Italian design, which has been acquired by the United Kingdom and many other countries. In a way it is a modern successor to pack and mountain artillery, and it can be dismantled easily for movement and re-assembled quickly. The various parts are small enough for it to be man-packed for short distances, and with mules the 'gun linft consists of eleven loads. It is easily transportable by aircrae' and can be dropped by parachute. It is towed in the British Army by a Land Rover with a long wheel base. The trail is split with long cranked legs, and, for towing, the rear sections of these legs fold up and lock. Though mainly intended for close support, it can be used as an anti-tank gun if required, the legs being laid flat and the wheels taken out and moved forward about eighteen inches into sockets on suspension arms. So arranged, the piece presents a very low profile and is thus easily concealed. The range of this piece is 11,564 yards and the maximum elevation is 65°. Ammunition is Q.F. semi-fixed with six normal charges and a seventh super charge, and the weight of the shell is 33 lb. There is a five-stage muzzle brake to take the extra recoil from the high velocity A.P. ammunition.[7]

The main range of British Army ordnance is completed by the 105-mm. light gun, British-designed, and intended in due course to replace the 105-mm. pack howitzer.[8] For towing, other than short distance movement, the barrel turns through 180° so that the muzzle is above the towing eye. There is a spring equilibrator on either side of the cradle to support the long barrel as on the 5.5-in. gun-howitzer, but these are mounted horizontally instead of vertically. It uses the same ammunition as the Abbot.[9]

In the French Army practically all artillery equipments are self-propelled. The staple field artillery weapon is the 105-mm. howitzer on the AMX 13 tank chassis, which has a maximum range of 16,500 yards and an elevation of 70°. The 155-mm. self-propelled howitzer is mounted on the same tank chassis, and, on account of this heavy loading, the trunnions are at the rear and the recoil force is transmitted direct to the two large spades at the back of the chassis. Range is 21,870 yards and the elevation 67°.[10]

As always, in the Russian Army, a very large proportion of its strength consists of artillery, but most of it is still towed. There is a very large 302-mm. (about 12-in.) gun-howitzer which has a range of 31,700 yards with a 300-lb shell. The medium equipment of the rifle division is the

23. A 25–pr gun-howitzer, showing the method of towing gun and limber.

24. Diagram of the 25–pr gun-howitzer showing recoil distances.

25.   A 5.5–in howitzer at Bou Arada, 1942.

26.   An American 105–mm self-propelled howitzer in action.
The gun is mounted on a Sherman Tank chassis.

155-mm. gun-howitzer, which has a range of 31,700 yards and fires a 132-lb shell. Its ammunition is separately loaded and charges are cased. The standard field equipment is the 122-mm. howitzer, which has been sold or presented to a number of communist and other countries. Some of these pieces have been mounted on tracked chassis for use as assault guns. The maximum range is 12,903 yards.[11]

Of tank guns, far the best is the 120-mm. B.L. high velocity piece of the British Chieftain tank. It has two types of projectile, Armour Piercing Discarded Sabot (APDS) and High Explosive Squash Head (HESH). The former has a high velocity and flat trajectory and is particularly effective against multiple armour plates. The latter has a thin wall and an H.E. filling and its main use is probably against troops in the open.[12] The comparable tanks of the major powers are the American M-60, the German Leopard, the French AMX 30, and the Russian T 55. The American and German tanks are armed with the British 105-mm., which, though an excellent gun, is much inferior to the 120-mm.; the French tanks carry a 105-mm. gun of their own design, and the Russian tanks have a 100-mm. gun of poorer performance.[13] The British 105-mm. gun is also the main armament of the Swedish STRV 103 (S Tank) and the Vickers 37-ton tank, which is the standard tank of the Indian Army under the name of Vijayanta.

At sea, guns are now very much overshadowed by missiles. The heaviest guns in the Royal Navy are the two 6-in. mounted in each of the helicopter cruisers *Tiger* and *Blake*. They are, of course, beaten handsomely by the four U.S. 'Iowa' class battleships, each armed with nine 16-in. guns; but these are obsolescent vessels. The standard British naval gun is the 4.5-in., whereas the U.S. Navy prefer the 5-in. The main armament of most ships is provided by the aircraft, the guided missile, or the torpedo.

CHAPTER XVIII

'Ammunition Without Ordnance'

T HE rocket, which in its modern form has superseded the ordnance-fired projectile as the principal artillery weapon, is the older missile of the two. The Chinese knew of it in very ancient days, though whether their rockets were other than fireworks is doubtful. Certainly the rocket can never have been a major weapon of war before the end of the eighteenth century, or its achievements would have been chronicled.

Early rockets consisted of a case made of wood, paper, or papier-mâché, filled with gunpowder, or some other suitable propellant, and bored out to suitable depth. Stability in flight was usually effected by a long stick fixed to and prolonging the case to the rear. That they were comparatively ineffective may be inferred from the fact that after the advent of ordnance they were little heard of in European warfare for some four hundred years.

War rockets seem to have been first brought to the notice of the Board of Ordnance through the effective use made of them against the troops of the Honourable East India Company at the siege of Seringapatam in 1799, during the war with Mysore. The Company's Adjutant-General reported that the troops suffered more from the enemy's rockets than from their guns.[1] This report led eventually to the introduction of rockets to the Royal Artillery. The Board of Ordnance asked the Royal Laboratory if they could provide an expert to initiate the manufacture of rockets. Having no such person, the Royal Laboratory suggested an approach to the East India Company, because the report had come from them. But neither could the East India Company give any help.[2]

Now the Comptroller of the Royal Laboratory, Woolwich, was General Sir William Congreve, Bart. (inventor of the block trail), who had been appointed to the post in 1793. His eldest son, Colonel William Congreve, was an officer in the Hanoverian Army and had himself been attached to the Royal Laboratory in 1791. Hearing (probably from his father) of this approach by the Board of Ordnance, he took up the study of rockets himself and soon became an enthusiast. The point which struck him immediately was that the rocket as a weapon required no ordnance for its discharge and had no recoil; it was, as he said, 'ammunition without ordnance'. Lacking official encouragement, he proceeded to carry out experiments at his own expense. At first the maximum range of his rockets was limited to 500 yards. He managed to increase this to 1,500 yards and then, with a 6-pr rocket, to 2,000 yards. So far all his experiments had been carried out with rockets having paper cases. In 1806 he changed over from a paper case to an iron case and used a shorter stick. A 32-pr rocket made to this pattern was 3 ft 6 in. long and 4 in. in diameter, with a 15-ft long stick and a maximum range of 3,000 yards.[3]

The first time that Congreve's rockets were used in action was in an attack on Boulogne harbour on 18 October, 1806, when 200 rockets were fired into the town from eighteen vessels and caused considerable destruction. In 1807 a similar attack was made against Copenhagen.[4] At the battle of Leipzig in 1813 a half troop of R.H.A. (which had been equipped with rockets), under the command of Captain Bogue, was the only British unit in the field. It was attached to the Army of the North under the command of the Crown Prince of Sweden, but had no opportunity of using its rockets in action until the battle of Leipzig, where they were a great success. Captain Bogue fell early on the day of the battle and command devolved on Lieutenant Fox Strangways. At the most critical period of the battle the Crown Prince of Sweden rode up to Strangways and implored him to advance his troop as nothing else could save the day. General Sir Edward Cust, the historian (1794–1878), in his account of the battle, says that such was the fearful effect of the rockets that a whole brigade surrendered after enduring their fire for a few minutes. Strangways received the personal thanks of the Allied Sovereigns, and the Emperor of Russia removed the Order of St Anne from his own breast and pinned it on Strangways.[5]

In February, 1814 a rocket battery played a significant part in the crossing of the Adour by the Peninsular Army. Of this episode Napier, in his *History of the War in the Peninsula and in the South of France*, says:

> On the night of the 22nd the first division, six eighteen-pounders, and the rocket battery, were cautiously filed from the causeway near the Anglet towards the Adour.... The French flotilla opened its fire about nine o'clock; Hope's artillery and rockets retorted so fiercely that three gun-boats were destroyed, and the sloop so hardly handled that the whole took refuge higher up the river; meanwhile sixty men of the guards were rowed in a pontoon across the mouth of the river in the face of a French piquet, which, seemingly bewildered, retired without firing. A raft being then formed with the remainder of the pontoons, a hawser was stretched across, and six hundred of the guards and the sixtieth regiment, with a part of the rocket battery, the whole under Colonel Stopford, passed. . . . Maucomble made a show of attacking Stopford, but the latter, flanked by the field artillery from the left bank, received him with a discharge of rockets; projectiles which, like elephants in ancient warfare, often turn upon their own side. This time, however, amenable to their directors, they smote the French column and it fled amazed, and with a loss of thirty wounded.

In 1814, Colonel William Congreve succeeded his father to the title and also as Comptroller of the Royal Laboratory. In the same year he published his little book, *Details of the Rocket System*. The opening words of the book are:

> His Royal Highness the Prince Regent, to whose gracious patronage the Rocket System owes its existence, having

been pleased to command the formation of a Rocket
Corps on the 1st of January, 1814, by augmentation to
the Regiment of Artillery, as proposed by his Lordship
the Earl of Mulgrave, Master General of the Ordnance;
I have thought it my duty to draw up the following
details of the System, for the instruction of the Officers
of the Corps, for the information of the General Officers
of the British Army, and that of such departments as it
is important for the good of the service, to make
acquainted with the principles of this new branch of our
naval and military means of offence and defence.

Congreve held that 'the very essence and spirit of the
Rocket System is the facility of firing a great number of
rounds in a short time, or even instantaneously, with small
means'. The rocket, indeed, was a kind of fixed ammunition
which could be fired without the need of ordnance and,
when an apparatus was used, it was a simple and easily
portable equipment. It followed that a mass of rockets could
be assembled quickly to deliver very powerful volleys.
In a defensive position sites should be prepared for rocket
batteries. One embrasure was needed for each rocket,
which could be made by excavating a channel in the
ground, to give direction, four to five feet long; such embrasures should be three feet apart. In battle, under conditions
where there was no time to prepare the ground in this
fashion, rockets would be discharged from the apparatus
and also, if it were sufficiently level, directly from the
ground. Against an advancing enemy it should be possible
to discharge a volley of up to 500 rockets which should
nearly annihilate the troops against which it was directed.
In the meantime the personnel of the battery should be well
protected, for, except for the one or two men who were
firing the volley, they could be kept under cover.
The maximum range and the height of the trajectory
depended on the length of the stick; the full-length stick
gave the longest range and the highest trajectory, while
the reduction of the length of the stick resulted in a flatter
trajectory and a shorter range. Firing from the ground, the

smaller rockets had ranges of 800 – 1,000 yards and the larger 1,000 – 1,200 yards.

In using the apparatus, the minimum angles necessary to get any longer range, as compared with firing from the ground, were 15° for the smaller rockets and 20–25° for the larger. Below these angles the rocket was apt to drop on leaving the apparatus and graze the ground immediately in front of it. In installing a rocket-bombarding battery the apparatus might be dispensed with by constructing an embankment or parapet at the required angle and of sufficient height for rocket and stick to be leaned against it. From such a battery, Congreve said, a volley of fifty rockets could be discharged every five minutes. He reckoned that twenty men were sufficient to maintain from a fixed position a rate of fire of a volley of twenty rounds of $5\frac{1}{2}$-in. howitzer shells, or 18 and 24 lb solid shot, three times a minute, and he invited comparison with the number of men required to keep up this rate with conventional artillery. He also pointed out that whereas the movement of ordnance required a considerable amount of transport, in a rocket battery there was only the carriage of ammunition to be considered. He made the further important point that there were types of ground over which ordnance could not be brought into action, whereas rockets could be taken wherever infantry could move.

When using the rocket apparatus, or frame, for high-angle shooting, the elevation had to be increased above that calculated, because a rocket did not obtain its full power immediately on leaving the frame and therefore dropped a few degrees at the start of its flight. The increase necessary was 5–10° for a large rocket and $2\frac{1}{2}$–5° for a small one. If there was a strong wind blowing against the rocket, it acted on the stick to depress the elevation, whereas a following wind had the reverse effect; the elevation on the frame had to be corrected accordingly. If there was a cross wind, adjustments had to be made in the opposite direction to those needed for conventional artillery, because the pressure on the stick tended to push the rocket to windward instead of to leeward. But it was only at high angles that any allowance had to be made for wind.

For firing the rocket from the ground, a fairly smooth and level surface was needed for the first 100 yards beyond the firing point, as rockets generally remained in contact with the ground for this distance. Beyond this range, power was sufficiently increased for rockets to surmount objects rising a few feet above the surface of the ground.

Congreve drew up the complete establishment for a troop of rocket horse artillery, with the same number of men and horses as there were in a normal horse artillery troop. The troop consisted of three divisions, each of two sub-divisions. A sub-division comprised five sections, each of three mounted troopers, and two mounted drivers, each leading two ammunition horses. Each mounted man carried into action four rounds of 12-pr rocket ammunition and each ammunition horse carried eighteen rounds. A section of three, therefore, carried twelve rounds (i.e., twelve rockets and twelve sticks) and one of the three had, in addition, a 'bouche-à-feu' – a small iron plate trough about 1 ft 6 in. long which was fixed to the ground by four iron points underneath it, and which was used for firing rockets from the ground.

In addition to this purely horsed establishment, Congreve proposed adding a two-horsed ammunition cart to carry three of the rockets assigned to each man, on the line of march to ease the burden on the horses. In action these carts would each carry 60 rounds of reserve ammunition, so that the sub-division would have a total of 200 rounds.

The complete establishment recommended by Congreve for the troop was as follows:

Officers	5
N.C.O.s	15
Troopers	90
Drivers	60
Artificers	8
Cars, heavy (each 4 men & 40 rounds 24-pr rockets)	3 (one per div.)
Cars, light (each 2 men & 60 rounds	

12-pr rockets)	3 (one per div.)
Curricle ammunition carts	6
Bouches-à-Feu	42
Ammunition heavy shell (24-pr), rounds	260
Ammunition light shell (12-pr), rounds	1200
Forge Cart	1
Horses	164

Mounted troopers carried their rockets in holsters, slung one on each side of the horse and supported by the pommel of the saddle. The 7-ft-long sticks were collected in a bundle, secured by a strap leading across the man's thigh to the peak of the saddle, and carried on the off side of the horse with the thicker ends resting in a bucket and the shafts falling naturally under the man's right arm.

On the ammunition horses the load was divided into three parts, a case of eighteen sticks carried on top and a saddle bag containing nine rockets on each side.

Congreve laid down a drill for a rocket sub-division coming into action. On the command, 'Prepare for action in front – dismount', Nos. 1 and 3 of each section dismounted and handed over their reins to No. 2 who remained mounted. No. 1 drew the bouche-à-feu from the leather case at the back of No. 2's valise and ran forward fifteen or twenty paces. Nos. 2 and 3 prepared a rocket drawn from any one of the holsters. No. 1 fixed the bouche-à-feu in the ground, pointing in the desired direction, and lit a portfire. No. 3 had by this time brought forward the first round and he laid it in the bouche-à-feu. On the command, 'Fire', No. 1 touched the vent of the rocket with his portfire. No. 3, in the meantime, had run back for another round, which No. 2 had been preparing. The rate of fire by the sub-division would be about two to three rounds per section per minute, or even faster with a well-trained troop.

Twelve light frames were held by each troop for firing rockets at high angles, and one of these was carried by each driver of ammunition horses. The frame was a tripod surmounted by an open cradle from which the rocket was fired. The cradle had an all-round traverse and could be elevated to any required angle.

On the command, 'Cease firing', No. 1 cut his portfire, took up the bouche-à-feu, and ran back to his section. No. 3 had already run back and had recovered the reins for No. 1 and himself. The sub-division could be on the move in less than a minute from the above command being given.

The heavy and light rocket cars given in the establishment consisted of a limber and carriage drawn by four horses. The heavy car was for the carriage of 24- or 32-pr rockets and the light car for 12- or 18-pr rockets. The former could carry 40 rounds of 24-pr and the latter 60 rounds of 12-pr or 50 rounds of 18-pr. The rockets were packed in boxes on the limber and the sticks were carried in half-lengths in boxes on the carriage, on which the men also rode on fixed seats. Between the boxes on the carriage was carried a double iron plate trough for the discharge of two rockets simultaneously. The trough could be either placed on the ground or attached to the carriage for high angle firing. For this latter purpose, the carriage was unlimbered, its perch detached, and the vehicle tipped forward on its axle in the direction of the target. The trough was then mounted in position, supported at the back by brackets attached to the carriage, and at the front by the up-ended perch. The whole was kept steady by a chain tightened between perch and axle-tree. Two men were required to fight the light car and four the heavy.

Congreve divided rockets into three categories – heavy, medium, and light. Heavy rockets were described by their diameter in inches, the medium rockets were those rated from 42-pr to 24-pr inclusive, and light rockets ran from 18-pr to 6-pr. The heavy rockets had a range of 2,000–2,500 yards, and their sticks were divided into four parts which were secured together by ferrules. The medium rockets had the longer range of 3,000 yards, were more portable than the heavy rockets, and were large enough for most bombardment tasks. Light rockets were far the most extensively used. The largest of them, the 18-pr, was armed with either shell or 9-pr solid shot. The 12-pr was armed with 6-pr shot, the 9-pr with a grenade, and the 6-pr with either shell or a 3-pr shot. Shells for use with rockets were made elliptical instead of round in order to decrease the air resistance. A percentage of all rockets between 9- and 24-pr were armed with case shot by putting a quantity of musket balls into a chamber in the top of the cylinder of the rocket, from whence they were discharged by powder of sufficient quantity to increase the velocity of the balls well beyond that of the rocket.

An external paper fuse was fitted to all rockets which it was intended should explode, and the fuse was ignited from the vent when the rocket was fired. The paper fuse, cut to the length needed, was connected to the bursting charge by a quick match contained in a tube fixed to the outside of the rocket.

All rocket sticks for land service were made into convenient lengths which were jointed together by iron ferrules, but for service at sea they were in one unbroken length. The rocket stick was attached to the rocket by three metal bands which surrounded the base of the rocket and the top end of the stick.

Congreve ends up his *Details of the Rocket System* as follows:

> It should be observed that, with due care, the Rocket ammunition is not only the most secure, but the most durable that can be: every Rocket is, in fact, a charge of powder hermetically sealed in a metal case, impervious either to the ordinary accidents by fire, or damage from humidity. I have used Rockets that had been three years on board of ship, without any apparent loss of power; and when after a certain period, which, from my present experience, I cannot estimate at less than eight or ten years, their force shall have so far suffered as to render them unserviceable, they may be again regenerated, at the mere expense of boring out the composition and re-driving it: the stick, case, &c., that is to say, all the principle parts, being as serviceable as ever.

Perhaps because they were never given sufficient trials to surmount their 'teething troubles', the Congreve rockets were never a complete success, and Napier's mention of them at the Adour, whilst doubtless exaggerating their defects, implies that they were disliked and distrusted by many soldiers. Wellington undoubtedly disliked them, and when a rocket detachment was sent out to him in the Peninsula in October, 1813 he received it with very mixed feelings, for, says Duncan, 'he had rather a horror of the rocket as a weapon of war'.[6] (It is a horror which would doubtless be echoed by very many people at the present time!)

Rockets were first added officially to the equipment of the Royal Artillery when, in January, 1813, 194 officers, N.C.O.'s, and men were added to the establishment of the Royal Horse Artillery to act as Rocket Detachments.[7] In 1814, the rocket detachments at home and overseas were organised respectively into the 1st and 2nd Rocket Troops. Captain W. G. Elliott was appointed to command the 1st Troop and Captain (later Sir) E. C. Whinyates was given command of the 2nd Troop. The 2nd Rocket Troop was present at the battle of Waterloo, but it nearly did not fight as such. At the beginning of May, 1815, the Duke of Wellington decided to change it into an ordinary troop of horse artillery, to the great distress of Whinyates, who loved his rockets. Wellington's official reason was that he was short of artillery, and when he had the proportion of ordnance that the armies of other nations had, he would welcome rockets; but he thought that for the operations in which he was engaged the gun was the superior weapon.[8] Sir George Wood, commanding the Artillery of the Army, persuaded the Duke to let Whinyates keep a proportion of 12-pr rockets with his guns. Actually, in the battle, Whinyates troop fired 309 shot, 236 spherical case, 15 common case, and 52 rockets[9] – figures which seem to support Wellington's opinion.

On 16 May, 1815 the following order was issued: 'His Royal Highness the Prince Regent, in the name and on behalf of His Majesty, has been pleased to command that

the Rocket Troop of Royal Artillery, which was present at
the Battle of Leipsic, be permitted to wear the word
'Leipsic' on their appointments, in commemoration of
their services on that occasion.' This same troop, the 2nd,
was also granted the honour 'Waterloo'. In 1816, however,
the 2nd Rocket Troop was disbanded and its honours
passed to the 1st Rocket Troop which had never been out
of England.[10] In 1847 the Rocket Troop became I Troop
of the Royal Horse Artillery and a rocket carriage was
then attached to every battery.[11] This decision was announced in a memorandum of 16 June, 1847 on 'Equipment for Rocket Artillery in the British Service' in which
it was stated:

> With a view of extending the knowledge of the Rocket
> Service in the Royal Horse Artillery, the Master-General
> has directed that each of the Troops in Great Britain
> shall have a Rocket Section as a part of its equipment,
> and that the Rocket Troop shall no longer form a
> distinct branch of the Service: the following alterations
> will therefore take place, viz.:
> The Rocket Troop becomes I. Troop, to be reduced in
> its number of horses to 41. C. and H. Troops to be
> supplied with a 6-pr. rocket carriage, &c., and their
> establishment to consist of 41 horses each.
> A Troop, on arrival from Ireland, to be completed in a
> like manner.
> To meet these arrangements, 12 horses are added to the
> establishment of the Horse Artillery.[12]

Rockets were still far from popular. The *Aide-Mémoire to
the Military Sciences* of 1852 says:

> Rocket Artillery is not in good repute as yet in any
> Service: we hear of its success in the late campaign in
> Lombardy and Hungary with the Austrian armies;
> but a partial success under favourable and judicious
> circumstances does not seem to tempt the military
> authorities to adopt this arm as a special Service for the
> Field.

There are peculiar difficulties in the working of the rocket, when influenced by extreme heat and cold, that have not yet been overcome: these contract the composition, and cause a void between it and the cases which produce explosion extremely dangerous.

It would appear that rockets left long in store are more easily affected by climate than those recently made.[13]

In the latter days of its use the Congreve rocket had undergone certain improvements. The iron case was fitted with two kinds of head, one which was used when the rocket served as either shell or solid shot, and the other when it was employed as a carcass (i.e., incendiary shell). The other end of the case was closed by an iron plate containing a threaded hole into which the stick was screwed (instead of being secured to the outside of the case). Around the hole for the stick were five vents in the plate for the escape of gas.

The shell head had an empty chamber which could be filled with powder through a hole which was closed by a metal plug. To ignite this powder there was a fuse connecting the chamber to the rocket composition which could be bored to the length required. The carcass head was pointed, filled with carcass composition, and had holes bored for the escape of the flame.

Congreve rockets were of four sizes (3-pr, 6-pr, 12-pr, and 24-pr), of which only the last two were supplied with carcass heads. They were packed in boxes containing, respectively, 30, 16, 9, and 6 rockets. Sticks for the rockets were tied up in bundles containing the same numbers.

The rocket carriage had, instead of a trail like a gun carriage, a long box for carrying the sticks. Two boxes for rockets were fixed to the sides of the carriage and there were two small boxes for implements and slow match. The tube from which the rockets were fired was carried on top and there was a box for horseshoes underneath. Some of the rockets, however, were carried in the limber boxes.[14]

In 1864 there were two types of rocket carriage, a 12-pr and a 6-pr. One carriage of either type accompanied each

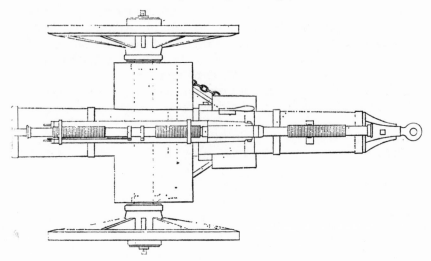

12–pr rocket wagon, 1852.

battery of field or horse artillery. The 12-pr had been essentially the field battery equipment and batteries of horse artillery had been equipped with 6-pr carriages, but the 12-pr was becoming the universal equipment, though the carriage carried less than half the number of rockets – 100 instead of 216.[15]

Colonel E. M. Boxer, R.A., inventor of many devices connected with artillery ammunition, produced rockets embodying improvements on Congreve's design which were adopted in September, 1864 to replace the latter's 6-pr, 12-pr and 24-pr types. The chief improvements were two: firstly, the position of the vents at the base of the rocket were altered to remove a weakness which had resulted in rockets bursting; and secondly, a stronger composition was incorporated which imparted a higher initial velocity to the rocket. In October, 1866 Boxer's 3-pr rocket was provisionally approved; in the preceding August all Congreve's rockets had been withdrawn.[16]

Meanwhile, William Hale, a mechanic in the Royal Arsenal, had invented a new type of rocket in 1845. The Hale rocket differed radically from the Congreve pattern, in that the flight was controlled by rotation instead of with a stick. The case of the rocket was, in its ultimate form, made of steel tubing. The cone-shaped head was cast iron, rivetted to the case, and separated from the composition by a disc of millboard. The composition (a mixture of saltpetre, sulphur and charcoal) was made up into pellets and pressed into the case hydraulically. A conically shaped hole was then bored through the composition for about two-thirds of its length. A washer, also of millboard, was

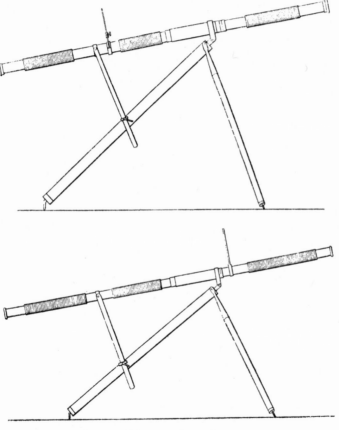

12–pr and 6–pr rocket tubes, 1852.

then pushed behind the composition and the case was closed by screwing on a wrought iron base piece. Into this in turn was screwed a tailpiece of cast iron which had three vents tapering towards the tail of the rocket, and so cut away on one side that the gas issuing through the vents imparted a thrust to the rocket and made it rotate in flight.[17]

Not receiving much encouragement in Great Britain, Hale took his invention to the United States and demonstrated it successfully at Washington, on 1 December, 1846, to officers of the Army and Navy. On 5 January, 1847 he gave another successful demonstration in the Washington Arsenal.[18] As a result the Hale rocket was adopted in the United States Army. There were two sizes: a 6-pr with an outside diameter of 2½ in., and a 16-pr with a diameter of 3¼ in. At 5° elevation the range was 500–600 yards and at 47° the small rocket had a range of 1,760 yards and the large one 2,200 yards. They were usually fired from tubes, supported at the front by two legs and having a tall backsight.[19] They were used in the American Civil War, but not greatly used as their flight was so erratic.

In the Civil War the Confederate States had some Congreve rockets. Lieutenant-Colonel W. W. Blackford, C.S.A. says that General Jeb Stuart:

opened on them with a Congreve rocket battery, the first and last time the latter ever appeared in action with us. It had been gotten up by some foreign chap who managed it on this occasion. They were huge rockets, fired from a sort of gun carriage, with a shell at the end which exploded in due time, scattering 'liquid damnation', as the men called it. Their course was erratic; they went straight enough in their first flight, but after striking, the flight might be continued in any other course, even directly back towards where it came from. Great consternation was occasioned among the camps of the enemy as these unearthly serpents went zigzagging about among them. . . . A few tents were fired but the rockets proved to be of little practical value.[20]

27. The American 155-mm howitzer 'Long Tom' being fired.

28. Rocket ammunition horse.

29. Bombardment Rockets.

In 1867 Hale's rockets were adopted in the United Kingdom in replacement of Boxer's, the same four sizes being retained. The 3-pr and 12-pr were soon discontinued, however, and the 6-pr was reclassified as a 9-pr, because this was in fact its actual weight.[21] The apparatus first adopted for launching Hale rockets was not a great success and in 1868 it was superseded by a much more effective trough machine. But rockets were now suffering from the competition of rifled ordnance, with its far greater accuracy and range, and supersession of them in their existing form was inevitable. Even the improved Hale rockets were slow and inaccurate in flight, variations in wind could change their direction, and alterations in their centre of gravity due to the consumption of fuel could affect their range; in addition, they deteriorated rapidly in storage.[22]

From the 1870s, rockets were little used, though succeeding Marks were introduced up to VII, and it was only after the First World War, on 11 September, 1919, that they were withdrawn from Service use.[23]

For the next fifteen years no further work was done in the United Kingdom on war rockets. Then in 1934 the new Master-General of the Ordnance (General Sir Hugh Elles, commander of the Tank Corps in the First World War), who was concerned by German progress on rocket flight, directed a review of the potential of war rockets. As a consequence of this review a Rocket Advisory Committee was set up in 1936 to advise on rocket development.

It was soon apparent that the weapon discarded as obsolete in 1919 had in fact been only in the early stages of its development and that Congreve might not after all have been entirely wrong in believing that he had invented a weapon which could rival ordnance. German progress was, indeed, alarming. In 1933 the Germans had built their A1 rocket which was $4\frac{1}{2}$ ft long, 1 ft in diameter, and weighed 330 lb, and in 1934 they improved on it with the A2 of practically the same dimensions but with more powerful propulsion. In 1938 there was a significant advance in the A3 rocket, which was 25 ft long, $2\frac{1}{2}$ ft in diameter, 1,650 lb in weight, and able to reach a range of 11 miles. Nevertheless, this was only a forerunner to the great A4 rocket (known in Great Britain as the V2) of 1940. This was 46 ft long by 5½ ft diameter and weighed 12½ tons. But perhaps its most striking characteristic was its then unheard-of range of 180–220 miles which, for the first time in history, took a missile out of the tactical and into the strategical field. In flight it reached a height of nearly 60 miles and a speed of some 3,500 m.p.h. This formidable weapon was driven by two fuels in separate tanks, liquid oxygen and a 75 per cent solution of ethyl alcohol and water. At the front of the rocket was the warhead and at the rear the main combustion chamber and exit nozzles. Control equipment cut off the fuel supply when the rocket reached the required range. It was launched vertically but it gradually deflected from the vertical till at about 22 miles up it was about 45° above the horizontal.[24] This was, of course, the genesis of both the evil inter-continental ballistic missile and the vehicle for extra-territorial flight.

British rocket development was in a more modest sphere and was confined to tactical use. After successful trials in Jamaica in 1939, rockets were manufactured in three sizes – 2 in., 3 in., and 5 in. Of these the 3-in. rocket was used for anti-aircraft defence in either a single or a twin mounting. It was 6 ft long and made of steel tubing. Its weight was 54 lb and it incorporated a H.E. shell, detonated by a fuse operated by air pressure. The propellant was a hollow cylinder of cordite. A battery equipped with these firing at German aircraft was a most impressive sight – the volley of rockets roared up into the air, apparently towing lines of red hot wires behind them. Some landing craft were fitted with multiple projectors called mattresses which launched 5-in. rockets and had a range of 3,800 yards. They could lay down a most devastating fire which made one glad not to be at the receiving end. A later equipment, the land mattress, with a range of 8,000 yards was used very effectively at the major river crossings in 1945. Aircraft also were fitted with rockets which were used both against submarines and in support of troops in the field.[25]

British rockets, then, were all tactical, whereas the Germans pioneered the long-range strategic weapons which, fitted with nuclear warheads, can never be used by any nation without committing national suicide.

The latest American and Russian inter-continental ballistic missiles show, by their very name, the extent to which long-range rockets have developed since the German V2, and the Polaris submarines bear witness to Great Britain's venture into the strategic rocket field. In the tactical sphere, with which this book is more concerned, the British Army acquired the American Corporal tactical atomic missile in 1959. This has a range of about seventy-five miles but it is a fairly static equipment.[26] A year later another American rocket came into service with the British Army, the Honest John. This is essentially a field artillery weapon, used in the same way as Congreve's rockets. The maximum range of the missile is about 24 miles. It is a free-flight rocket with no electrical controls and it is as accurate as contemporary ordnance and very mobile. The warhead is either H.E. or nuclear.

With somewhat similar range, the British Seaslug is the main armament of the 'County' class destroyers and was fitted to the first of them in 1961. It is primarily an anti-aircraft weapon but it can also be used against surface targets. The system tracks the target automatically.[27] This missile makes the County class destroyer the most formidable surface fighting ship that the Royal Navy has ever possessed. In spite of being classed as a destroyer, she could destroy any gun-armed battleship that ever existed.

Seacat is a smaller close-range anti-aircraft rocket which can be used also against surface vessels within sight. It can perhaps be regarded as the rocket equivalent of the duel purpose 4.5-in. gun. It has a H.E. warhead and its maximum range is about 3,500 yards. The operator keeps his sights on the target during flight and a command radio link keeps the missile on course.[28]

The main Royal Air Force and Army anti-aircraft weapons are Bloodhound and Thunderbird respectively. The Mark I Bloodhound system went into service in 1958 and the issue of Mark II started in 1964. Associated with a Bloodhound battery of four launchers is a target illuminated radar (TIR). From the information supplied by this, the missile homes on the target.[29]

The Thunderbird Mark I was issued to the Army in 1960, and the Mark II in 1965. It operates on the same system as Bloodhound.[30]

Two notable British anti-tank rockets are Vigilant and Swingfire. Vigilant, after test firings in 1957-8, went into production in 1960 and has been sold to several foreign countries. It is a man-portable wire-guided missile for use by infantry or in light armoured fighting vehicles, and is effective against the most heavily armoured tank. The system is put into operation by setting a launcher box on the ground, with the missile already inserted, and pointing it in the direction of the desired target. The firer has a sight-controller with a pocket battery and a length of connecting cable, which permits him to take up a suitable position up to about 65 yards from the launcher box. By pressing the firing trigger the missile is launched towards the target, trailing a wire behind it and emitting a flare to show its position. By moving the guidance control cup the firer can alter the direction of the missile and so guide it on to the target. (One can picture Congreve's delight at this refinement.) The maximum range is about 1,400 yards.[31]

Swingfire is a later wire-guided missile system with a much longer maximum range of over 3,000 yards and designed primarily for firing from a vehicle. The system went into operation with the British Army in 1969. In this system also the firer can be some distance away from the launcher so that he can be concealed behind cover.[32]

Perhaps the most controversial of land surface-to-surface missile systems is the American Shillelagh. It is fired from a combined gun-launcher and is primarily intended for mounting in armoured fighting vehicles. The 155-mm. gun-launcher will also fire conventional ammunition. When firing rockets, the gunner guides the missile to the target by an infra-red transmitter working to a receiver on the missile, and has merely to keep his sights trained on the target.

Shillelagh is the main armament of the Sheridan light tank. It is also intended for use by the United States and West German Armies in the projected main battle tank MBT 70. Progress in design however, has not been happy. In any case the Germans, if they take the tank, prefer a 120-mm. gun to the Shillelagh. The reason for this is that tests have shown the Shillelagh to be little more accurate than ordnance-fired projectiles. In addition, Shillelagh is subject to interference with its guidance system, is far more expensive than a shell, and the gunner has to hold his sights accurately on the target during the flight, which would require very strong nerves under fire.[33] Perhaps Sir William Congreve has not yet had the last word.

REFERENCES

Chapter I

1. F. L. Robertson, *The Evolution of Naval Armament* (1921).
2. B. St. J. O'Niel, *Castles and Cannon* (1960).
3. Robertson, op. cit.
4. Sir Charles Oman, *A History of the Art of War in the Middle Ages* (1924).
5. Robertson, op. cit.
6. Brigadier O. F. G. Hogg, *English Artillery 1326–1716* (1970).
7. Robertson, op. cit.
8. O'Niel, op. cit.
9. Oman, op. cit.
10. Robertson, op. cit.
11. O'Niel, op. cit.
12. Ibid.
13. Oman, op. cit., O'Niel, op. cit.
14. Hogg, op. cit.
15. Ibid.
16. Oman, op. cit.
17. Ibid.
18. Ibid.
19. Ibid.
20. O'Niel, op. cit.
21. Colonel Sir Bruce Seton, *The Flodden Campaign 1513. Journal of the Society for Army Historical Research, Volume III.*
22. O'Niel, op. cit.
23. Ibid.
24. Ibid.
25. Ibid.
26. Hogg, op. cit.

Chapter II

1. F. L. Robertson, *The Evolution of Naval Armament* (1921).
2. Ibid.
3. Ibid.
4. Colonel Sir Bruce Seton, *The Flodden Campaign 1513. Journal of the Society for Army Historical Research, Volume III.*
5. Ibid.
6. Ibid.
7. Ibid.
8. Robertson, op. cit.
9. Seton, op. cit.
   Brigadier O. F. G. Hogg, *English Artillery, 1326–1716* (1970).
10. Hogg, op. cit.
    E. W. Lloyd and Sir A. G. Hadcock, *Artillery: Its Progress and Present Position* (1893).
11. E. H. H. Archibald, *The Wooden Fighting Ship in the Royal Navy* (1968).
12. Sir Charles Oman, *A History of the Art of War in the Middle Ages* (1924).
13. B. St. J. O'Niel, *Castles and Cannon* (1960).

References

14. Oman, op. cit.
15. Ibid.
16. A. W. Wilson, *The Story of the Gun* (1944).
17. O'Niel, op. cit.
18. Oman, *A History of the Art of War in the Sixteenth Century* (1937).
19. O'Niel, op. cit.
20. Oman, *A History of the Art of War in the Middle Ages* (1924).
21. W. H. Prescott, *History of the Reign of Ferdinand and Isabella* (ed. J. F. Kuk 1887).
22. Major-General J. F. C. Fuller, *The Decisive Battles of the Western World* (1954–6).

Chapter III

1. Sir Charles Oman, *A History of the Art of War in the Sixteenth Century* (1937).
2. 'The Loyal Serviteur', *History of Bayard* (English Translation 1883).
3. Brigadier O. F. G. Hogg, *English Artillery, 1326–1716* (1970).
   B. St. J. O'Niel, *Castles and Cannon* (1960).
4. F. L. Robertson, *The Evolution of Naval Armament* (1921).
5. O'Niel, op. cit.
6. Hogg, op. cit.
7. O'Niel, op. cit.
8. Ibid.
9. Colonel Sir Bruce Seton, *The Flodden Campaign 1513. Journal of the Society for Army Historical Research, Volume III.*
10. Ibid.
11. Ibid.
12. E. W. Lloyd and Sir A. G. Hadcock, *Artillery: Its Progress and Present Position* (1893).
13. Robertson, op. cit.
14. Ibid.
15. John Muller, *A Treatise of Artillery* (1780).
16. Hogg, op. cit.
17. Oman, op. cit.
18. Robertson, op. cit.
19. Ibid.
20. Ibid.
21. E. H. H. Archibald, *The Wooden Fighting Ship in the Royal Navy* (1968).
22. Robertson, op. cit.
23. Lindsay Boynton, *The Elizabethan Militia* (1967).
24. Ibid.

Chapter IV

1. Colonel F. A. Whinyates, *From Corunna to Sevastopol* (1884).
2. Major-General J. F. C. Fuller, *The Decisive Battles of the Western World* (1954–6).
3. Ibid.
4. Ibid.
5. F. L. Robertson, *The Evolution of Naval Armament* (1921).
6. Colonel H. C. B. Rogers, *Battles and Generals of the Civil Wars 1642–1651* (1968).
7. Ibid.
8. Serjeant-Major Richard Elton, *The Compleat Body of the Art Military* (1650), 598 i. 21.

References

9. Brigadier O. F. G. Hogg, *English Artillery, 1326–1716* (1970).
10. Major Frederick Miller, R.A., V.C., *Equipment of Artillery* (1864–5).
11. Major-General G. G. Lewis & Others, *Aide-Memoire to the Military Sciences* (1853).
12. E. H. H. Archibald, *The Wooden Fighting Ship in the Royal Navy* (1968).
13. Robertson, op. cit.
14. Miller, op. cit.

CHAPTER V

1. Colonel F. A. Whinyates, *From Corunna to Sevastopol* (1884).
2. Ibid.
3. F. L. Robertson, *The Evolution of Naval Armament* (1921).
   Captain de Riviers de Mauny, *Explanations to Notes on Artillery dictated by Napoleon at St. Helens to Baron Gourgaud.*
4. Whinyates, op. cit.
5. Robertson, op. cit.

CHAPTER VI

1. E. W. Lloyd and Sir A. G. Hadcock, *Artillery: its Progress and Present Position* (1893).
2. F. L. Robertson, *The Evolution of Naval Armament* (1921).
3. Ibid.
4. Ibid.
   Lloyd and Hadcock, op. cit.
   Major-General B. P. Hughes, *British Smooth-Bore Artillery* (1969).
5. John Muller, *A Treatise of Artillery* (1780).
6. Ibid.
7. Ibid.
8. The gauge of the Stockton & Darlington Railway, the first public railway in the world, was 4 ft 8 in. Another half-inch was later added by some person, so that the standard gauge was ultimately 4 ft 8½ in.
9. Hughes, op. cit.

CHAPTER VII

1. Colonel F. A. Whinyates, *From Corunna to Sevastopol* (1884).
2. Lieutenant A. W. Wilson, R.A., *The Story of the Gun* (2nd edition 1965).
3. Major F. Miller, V.C., R.A., *Equipment of Artillery* (1864–5).
4. Ibid.
5. Miller, op. cit.
6. Colonel Julian R. John Jocelyn, *The History of the Royal Artillery (Crimean Period)* (1911).
7. Miller, op. cit.
8. Jocelyn, op. cit.
9. Major-General G. G. Lewis & Others, *Aide-Memoire to the Military Sciences* (1853).
10. Major-General B. P. Hughes, *British Smooth-Bore Artillery* (1969).
11. Captain F. A. Griffiths, R.A., *The Artillerist's Manual* (1839).
12. Jocelyn, op. cit.

216

13. Miller, op. cit.
    Hughes, op. cit.
14. Captain J. F. Owen, R.A., *Treatise on the Construction of Ordnance* (1877).
15. Ibid.
16. Ibid.
17. F. L. Robertson, *The Evolution of Naval Armament* (1921).
18. Hughes, op. cit.
19. Ibid.
20. Ibid.
    E. W. Lloyd and A. G. Hadcock, *Artillery, its Progress and Present Position* (1893).
    Brigadier-General C. A. L. Graham, *The History of the Indian Mountain Artillery* (1957).
21. Ibid.
22. Hughes, op. cit.
23. Griffiths, op. cit.
24. Robertson, op. cit.
    Hughes, op. cit.
25. Brigadier O. F. G. Hogg; letter to Major G. Tylden.
26. Ibid.
27. Hughes, op. cit.
28. Hogg, op. cit.
    Jocelyn, op. cit.
    Miller, op. cit.
29. Hughes, op. cit.
30. Ibid.
    Wilson, op. cit.

CHAPTER VIII

1. Captain J. F. Owen, *Construction and Manufacture of Ordnance* (1877).
2. Ibid.
   F. L. Robertson, *The Evolution of Naval Armament* (1921).
3. E. W. Lloyd and Sir A. G. Hadcock, *Artillery: Its Progress and Present Position* (1893).
4. Ibid.
5. Ibid.
6. Owen, op. cit.
7. Robertson, op. cit.
8. Owen, op. cit.
   Robertson, op. cit.
9. Owen, op. cit.
   Lloyd & Hadcock, op. cit.
10. Robertson, op. cit.
11. Owen, op. cit.
12. Ibid.
13. Robertson, op. cit.
14. Owen, op. cit.
15. Ibid.
16. Ibid.
17. Ibid.

18. Ibid.
    Lieut. A. W. Wilson, R.A., *The Story of the Gun* (2nd edition 1965).
19. Ibid.
20. Owen, op. cit.
21. Lloyd & Hadcock, op. cit. (p. 297).
22. Ibid.
23. Ibid.
24. Brig-General C. A. L. Graham, *The History of the Indian Mountain Artillery* (1957).
25. Lloyd & Hadcock, op. cit.
26. Wilson, op. cit.
27. Owen, op. cit.
28. Ibid.
29. Ibid.
30. Ibid.
31. Ibid.
32. Lloyd & Hadcock, op. cit.
33. Lieut G. Will & Lieut J. C. Dalton, *The Artillerist's Hand-book of Reference* (1876).
34. Wilson, op. cit.
35. Owen, op. cit.

Chapter IX

1. Oscar Parkes, *British Battleships* (1957).
2. E. W. Lloyd & A. G. Hadcock, *Artillery: Its Progress and Present Position* (1893).
3. Ibid.
4. Ibid.
5. Lieut A. W. Wilson, R.A., *The Story of the Gun* (2nd edition, 1965).
6. Lloyd & Hadcock, op. cit.
7. Ibid.
8. Ibid.
9. Wilson, op. cit.
10. Ibid.
11. Lloyd & Hadcock, op. cit.
12. Ibid.
13. Wilson, op. cit.
14. Lloyd & Hadcock, op. cit.
15. Ibid.
    Oscar Parkes, op. cit.
16. Lloyd & Hadcock, op. cit.
17. Ibid.
18. Ibid.
19. Wilson, op. cit.
20. Ibid.

Chapter X

1. Major-General Sir Stanley von Donop, *Artillery Equipments used in the Field during the last 25 Years.*
2. *Treatise on Military Carriages* (1911, Official Publication).

References

3. Von Donop, op. cit.
4. *Treatise on Military Carriages* (1911).
5. Ibid.
6. Ibid.
7. Ibid.
8. Lieutenant A. W. Wilson, R.A., *The Story of the Gun* (2nd edition 1965).
9. Ibid.
10. Colonel H. A. Bethell, R.F.A., *Modern Guns and Gunnery* (1910).
11. Michael Glover, quoting *Letters of Sir Augustus Frazer* in *Wellington as a Military Commander* (1968).
12. Bethell, op. cit.
13. Von Donop, op. cit.
14. Ibid.
15. Brigadier-General C. A. L. Graham, *The History of the Indian Mountain Artillery* (1957).
16. Ibid.
17. Ibid.
18. Ibid.
19. *Treatise on Military Carriages* (1911).
20. Ibid.

Chapter XI

1. Lieutenant A. W. Wilson, R.A., *The Story of the Gun* (2nd edition, 1965).
2. Ibid.
3. Colonel H. A. Bethell, R.A., *Modern Guns and Gunnery* (1910).
4. Ibid.
5. Major-General Sir Stanley von Donop, *Artillery Equipments used in the Field during the last 25 Years*.
6. Von Donop, op. cit.
7. Bethell, op. cit.
8. Ibid.
9. Ibid.
10. *Treatise on Military Carriages* (1911).
11. Von Donop, op. cit.
12. Bethell, op. cit.
13. Ibid.
14. Ibid.
15. Ibid.
16. Wilson, op. cit.
17. Ibid.
18. Von Donop, op. cit.
19. *Military Carriages*, op cit.
20. Ibid.

Chapter XII

1. Oscar Parkes, *British Battleships* (1957).
2. Ibid.
3. Ibid.
4. Ibid.

5. Ibid.
6. Ibid.
7. Ibid.
8. Ibid.
9. F. Robertson, *The Evolution of Naval Armament* (1921).
10. Oscar Parkes, op. cit.
11. Robertson, op. cit.
12. Ibid.
    Oscar Parkes, op. cit.
13. Robertson, op. cit.
14. A. Temple Patterson, *The Jellicoe Papers* (Vol. I, Navy Record Society, 1966).
15. Ibid.
16. Arthur J. Marder, *From the Dreadnought to Scapa Flow* (1964).
17. Edgar J. Marsh, *British Destroyers* (1966).

Chapter XIII

1. Lt. Colonel Sir Reginald Rankin, Bt., *The Inner History of the Balkan War* (1914).
2. Ibid.
3. *Notes on the Balkan Wars (War Office, 1914).*
4. Alain de Pennenrun, *40 Jours de Guerre dans les Balkans* (1914).
5. *Notes,* op. cit.
6. Ibid.
7. Ibid.
8. Rankin, op. cit.
9. Major P. Howell, *The Campaign in Thrace* (1912).
10. Ibid.
11. General Izzet Fuad Pasha, *Paroles de Vaincu* (1914).
12. General Mahmoud Mukhtar Pasha, *Mon Commandement au Cours de la Campagne des Balkans de 1912* (1914).
13. Ibid.
14. Ibid.
15. Ibid.
16. Ibid.

Chapter XIV

1. Major-General Sir Stanley von Donop, *Artillery Equipments used in the Field during the last 25 Years.*
2. Ibid.
3. Mark Severn, *The Gambardier* (1930).
4. Ibid.
   Von Donop, op. cit.
   Treatise on Military Carriages (1911).
5. Severn, op. cit.
   Von Donop, op. cit.
   Military Carriages, op. cit.
6. Severn, op. cit.
   Von Donop, op. cit.
   Military Carriages, op. cit.
7. Von Donop, op. cit.
   Severn, op. cit.

References

8. Von Donop, op. cit.
   Severn, op. cit.
9. Von Donop, op. cit.
   Severn, op. cit.
10. Von Donop, op. cit.
    Severn, op. cit.
11. Von Donop, op. cit.
    Severn, op. cit.
12. Colonel H. C. B. Rogers, *Tanks in Battle* (1965).
13. Ibid.
14. Brigadier-General Sir James Edmonds, *A Short History of World War I*.
15. Severn, op. cit.
16. Edmonds, op. cit.
17. Ibid.
18. Ibid.
19. Rogers, op. cit.
20. Severn, op. cit.
21. Brigadier-General C. A. L. Graham, *The History of the Indian Mountain Artillery* (1957).
22. Ibid.
    Sir Patrick Cadell, *The History of the Bombay Army* (1938).
23. Graham, op. cit.
24. A. Temple Patterson, *The Jellicoe Papers* (Vol. I, 1966).
25. Ibid.
26. Patterson, op. cit.
27. Arthur J. Marder, *From the Dreadnought to Scapa Flow* (1966).
28. Patterson, op. cit.
29. Ibid.
30. Ibid.

CHAPTER XV

1. Richard M. Ogorkiewicz, *Armour* (1960).
   Lieutenant A. W. Wilson, *The Story of the Gun* (2nd edition, 1965).
2. Wilson, op. cit.
3. Ogorkiewicz, op. cit.
4. Wilson, op. cit.
5. Colonel H. A. Bethell, *Modern Guns and Gunnery* (1910).
6. Brigadier O. F. G. Hogg, *Artillery): Its Origin, Heyday and Decline* (1970).
7. Wilson, op. cit.
8. Hogg, op. cit.
9. Wilson, op. cit.
10. Hogg, op. cit.
11. Wilson, op. cit.
12. Ogorkiewicz, op. cit.
13. Hogg, op. cit.
14. Ibid.
15. Wilson, op. cit.
16. Hogg, op. cit.
17. Ibid.
18. Ibid.
    Wilson, op. cit.

19. Colonel H. C. B. Rogers, *Tanks in Battle* (1965).
20. Brigadier-General C. A. L. Graham, *The History of the Indian Mountain Artillery* (1957).
21. Oscar Parkes, *British Battleships* (1957).

Chapter XVI

1. M. M. Postan, *British War Production* (History of the Second World War) (1952).
2. Ibid.
3. Ibid.
4. Ibid.
5. Colonel H. C. B. Rogers, *Tanks in Battle* (1965).
6. Jane's *Weapon Systems, 1969–70* (Edited by R. T. Pretty and D. H. R. Archer).
7. Rogers, op. cit.
8. Ibid.
9. Ibid.
10. Ibid.
11. Ibid.
12. Ibid.
13. Richard M. Ogorkiewicz, *Armour* (1960).
14. Ibid.
15. Ibid.
16. Ibid.
17. Brigadier-General C. A. L. Graham, *The History of the Indian Mountain Artillery* (1957).
18. Vice-Admiral Sir Arthur Hezlet, *Aircraft & Sea Power* (1970).
19. Oscar Parkes, *British Battleships* (1957).
20. Edgar J. March, *British Destroyers* (1966).
21. Ibid.

Chapter XVII

1. Stevenson Pugh, *Fighting Vehicles and Weapons of the Modern British Army* (1962).
2. Jane's *Weapon Systems, 1969–70.*
3. Pugh, op. cit.
4. Ibid.
5. Jane's, op. cit.
6. Ibid.
7. Ibid.
   Pugh, op. cit.
8. Hugh Howton, *Soldier* (July, 1970).
9. Jane's op. cit.
10. Ibid.
11. Ibid.
12. Interavia, *The Modern Way – Weapons and Technology* (No. 2/1965).
13. Ibid.

Chapter XVIII

1. Brigadier O. F. G. Hogg, *Artillery: its Origin, Heyday and Decline* (1970).
2. Ibid.

References

3. Ibid.
4. Ibid.
5. Captain Francis Duncan, *History of the Royal Regiment of Artillery* (1873).
6. Ibid.
7. Ibid.
8. Ibid.
9. Ibid.
10. Ibid.
11. Ibid.
    Major Miller, V.C., *Equipment of Artillery* (1864).
12. Major-General G. G. Lewis and Others, *Aide Memoire to the Military Sciences* (2nd edition, 1853).
13. Ibid.
14. Miller, op. cit.
15. Ibid.
16. Hogg, op. cit.
17. Ibid.
18. Ibid.
19. Jack Coggins, *Arms and Equipment of the Civil War* (1962).
20. Lt. Colonel W. W. Blackford, *War Years with Jeb Stuart.*
21. Hogg, op. cit.
22. Ibid.
23. Ibid.
24. Ibid.
25. Ibid.
26. Stevenson Pugh, *Fighting Vehicles and Weapons of the Modern British Army* (1962).
27. Jane's *Weapon Systems, 1969–70.*
28. Ibid.
29. Ibid.
30. Ibid.
31. Ibid.
32. Ibid.
33. Ibid.

INDEX